W0106995

Progress in Colloid & Polymer Science

Vol. 94 (1994)

PROGRESS IN COLLOID & POLYMER SCIENCE

Editors: F. Kremer (Leipzig) and G. Lagaly (Kiel)

Volume 94 (1994)

Ultracentrifugation

Guest Editor:

M. D. Lechner (Osnabrück)

Springer-Verlag Berlin Heidelberg GmbH

Die Deutsche Bibliothek – CIP-Einheitsaufnahme

Ultracentrifugation / guest ed.: M. D. Lechner. – Darmstadt :
Steinkopff ; New York : Springer, 1994
  (Progress in colloid & polymer science ; Vol. 94)

NE: Lechner, Manfred D. [Hrsg.]; GT

ISSN 0340-255 X

This work is subject to copyright. All rights are
reserved, whether the whole or part of the
material is concerned, specifically those rights
of translation, reprinting, reuse of illustra-
tions, recitation, broadcasting, reproduction
on microfilms or in other ways, and storage
in data banks. Duplication of this publication
or parts thereof is only permitted under the
provisions of the German Copyright Law of
September 9, 1965, in its version of June 24,
1985, and a copyright fee must always be paid.
Violations fall under the prosecution act of the
German Copyright Law.

The use of registered names, trademarks, etc.
in this publication does not imply, even in the
absence of specific statement, that such names
are exempt from the relevant protective laws
and regulations and therefore free for general
use.

© Springer-Verlag Berlin Heidelberg 1994
Originally published by Dr. Dietrich Steinkopff Verlag
 GmbH & Co. KG, Darmstadt in 1994
Softcover reprint of the hardcover 1st edition 1994

Chemistry editor: Dr. Maria Magdalene
Nabbe; English editor: James C. Willis;
Production: Holger Frey, Bärbel Flauaus.

Type-Setting: Macmillan India Ltd.
Bangalore (India)

ISBN 978-3-662-15672-8   ISBN 978-3-7985-1675-5 (eBook)
DOI 10.1007/978-3-7985-1675-5

Preface

The symposium on analytical ultracentrifugation (AUC) is now well established since its initiation in 1978; the meetings take place at different universities every two years. The aim of all eight symposia was to bring together scientists who are working in the various fields of ultracentrifugation, including technical developments, applications, general theory, and results of investigation.

The centers of interest of the 8th symposium included biochemical and biophysical applications of AUC followed by characterization of polyelectrolytes, synthetic polymers, gels and latices. Additionally, the new, modern, fully computerized analytical ultracentrifuge Optima XL-A was presented by representatives from Beckman Instr. and an outlook for future technical developments was given by T. Laue (New Hampshire, USA). The symposium ended with a very stimulating discussion on the future of the analytical ultracentrifuge, new developments, cooperation of the user groups, and about the next symposium which will be held in spring 1995 in one of the new German states.

The 8th symposium was kindly sponsored by BASF AG (Ludwigshafen), Beckman Instr. (Munich), Kämmerer GmbH (Osnabrück), Kromschröder AG (Osnabrück), Röhm GmbH (Darmstadt), and Universitäts-Gesellschaft (Osnabrück). Support of this volume by the editor of the journal Colloid Polymer Science, H.-G. Kilian, is gratefully acknowledged.

M. D. Lechner (Osnabrück)

Contents

Progress in Colloid & Polymer Science   Progr Colloid Polym Sci 94:1–13 (1994)

Simultaneous radial and wavelength analysis with the Optima XL-A analytical ultracentrifuge

P. Schuck

Institut für Biophysik der Johann Wolfgang Goethe-Universität and Max-Planck-Institut für Biophysik, Frankfurt am Main, FRG

Abstract: We exploit the possibility offered by the new analytical ultracentrifuge Beckman Optima XL-A to record wavelength profiles at different radial positions and radial profiles at different wavelengths in sedimentation equilibrium experiments to unravel complexes formed by entities with different absorption spectra. The mathematical analysis is described here in detail. With a little *a priori* knowledge about the absorption spectra it is possible to resolve a greater number of components than with conventional methods. This method also enables one to achieve a simultaneous detection of the components' extinction profiles. In addition, an analysis of each component's statistical accuracy is presented.

Key words: Analytical ultracentrifugation – sedimentation equilibrium – heterogeneous associations – exponential fitting

1. Introduction

Analytical ultracentrifugation is an important tool to investigate interactions of biological macromolecules. In particular, weak reversible interactions can be characterized conveniently by sedimentation equilibrium experiments without disturbing association equilibria [1, 2]. Research on structure and function of supramolecular assemblies, such as, for example, multienzyme complexes, is a field that is steadily growing in importance. Therefore, it is necessary to study the behavior of many macromolecules and/or ligands in association equilibrium. This is probably the major task for present and future analytical ultracentrifugation.

Sedimentation equilibrium experiments have until now an inherent limitation in the complexity of the system that can be investigated. The reason is that the analysis of the measured radial absorbance profile requires a decomposition into Boltzmann exponentials of each component. Unfortunately, the unravelling of exponentials is one of the most difficult problems in applied data

analysis, and it is well-known that the parameters sought for the exponentials are badly determined by the data [3]. Therefore, in studies on multi-component systems, relative molar masses and partial specific volumes are mainly determined in separate experiments or amino acid analysis. In the present paper, this will be assumed unless indicated otherwise. Nevertheless, the evaluation of the radial profiles is ill-conditioned if more than three or four components are present in the solution [1]. In this case, many completely different sums of the exponentials will fit the data almost equally well, and no reliable information on the solutes is obtainable.

The analytical ultracentrifuge provides specific means to circumvent the problems just described. In the early 1960s, one of the primary motivations [4] for Schachman's pioneering work in the development of the photoelectric scanner with monochromator [4, 5] was the capacity to discriminate the components by varying the wavelength of the monochromator. This additional dimension in data space enables the analysis of more complex systems if the species under investigation have different absorption spectra, a situation which may be created by a selective labeling of species [2, 6–8].

One approach, proposed by Schachman [5], is to scan the radial profile at a particular wavelength where only one species absorbs [6–9]. Although this method eliminates the need to consider the other species during the analysis, it is very difficult to determine the association constants for interactions in which "invisible" species are involved [10]. It is also not applicable if the absorption spectra overlap and, therefore, it is restricted to special cases. A more general approach is the simultaneous analysis of two radial profiles scanned at different wavelengths [10–12]. This improves the conditioning of the analysis. But the results depend sensitively on the values of two extinction coefficients for each component that have to be used as prior knowledge.

Another possibility is to exploit a new source of information: the absorption spectra of the solution at different positions in the centrifugal field. They are available in the new Beckman Optima XL-A ultracentrifuge, and can be conveniently acquired in sedimentation equilibrium. A combination of absorption spectra scanned at multiple radii and radial profiles scanned at multiple wavelengths forms a two-dimensional data surface in a radius-wavelength-absorption space, which obviously contains a lot more information than single radial profiles. The analysis consists of a decomposition in terms of products of extinction profiles and exponentials, a problem that is generally much better conditioned than fitting exclusively with exponentials. Additionally, this two-dimensional data set contains information about the absorption spectra of all components. At least in principle this should offer possibilities to monitor spectral changes at distinct levels of interaction. It is the purpose of the present paper to investigate the information obtainable from such an extended data set, especially the number of components that can be resolved reliably with regard to their radial concentration distribution and their extinction profiles.

The data analysis proposed follows a three-step process. First, estimations of the components' extinction profiles will be calculated without considering the radial distribution in the centrifugal field. Second, with known extinction profiles,

the radial distribution will be evaluated. A third step combines the knowledge of the radial distribution and the extinction profiles to compute better approximations of the species extinction profiles, optionally constrained with different models of similarity between components. Steps two and three will be repeated until no further improvement of the fit occurs. Overinterpretation can be avoided by combining steps two and three with an analysis of the statistical accuracy of the results.

To test this method, it was applied to mixtures of the following components in detergent solution, each one with a different absorption spectrum: 1) An intrinsic membrane protein, the band 3 protein of the erythrocyte membrane, which occurs in a monomer/dimer/tetramer association equilibrium (relative molar mass of the monomer: $M_r = 100\,000$) [13]; 2) a small ligand, DBDS (4,4'-dibenzoamidostilbene-2,2'-disulfonate), reversibly associating with band 3 [14]; and 3) oxyhemoglobin (HbO), which also associates reversibly with band 3 [9] and which is in a dimer/tetramer association equilibrium. The systems studied in sedimentation equilibrium thus contained up to three different chromophores and 11 components.

2. Methods

2.1 Analysis of the statistical accuracy

When fitting exponentials to experimental data, it is essential to analyze the stability of the solution (dependence on noise). Methods to estimate confidence intervals in connection with nonlinear fits of ultacentrifuge data are established [15, 16]. However, for ideal solutions of species with known molecular weights and partial specific volumes, an efficient and exact method to analyze the statistical accuracy and to determine confidence intervals can be applied, if the problem to fit positive sums of Boltzmann exponentials to experimental absorbance values is formulated linearly and inequality-constrained. The sought parameters then are the concentrations at an arbitrary reference radius [1]. Their optimal values (and subsequently the association constants) can be calculated straightforwardly; non-negativity can be easily achieved as described in [17].

In general, the statistical accuracy of solutions to linear least-squares problems with non-negative parameters can be analyzed for each parameter k by tracing the goodness of fit parameter

$$\left(\; \chi_r^2(x) = \frac{1}{m-n} \sum_{i=1}^m \frac{1}{\sigma_i^2} \left[a(r_i) - \sum_{k=1}^n b_k(r_i) x_k \right]^2 \; 1 \right)$$

(where $x \in \mathfrak{R}^n$ is a vector with unknown parameters $(x_1, \ldots, x_n)$ with $x_j \geq 0$; $b_k(r)$ the basis functions to be fitted to the m observations $a(r_i)$ with standard deviations σ_i) in dependence on a constraint in the parameter x_k, while all other parameters are optimized:

$$x_k = c$$

$$\chi_{r,k}^2(c) = \frac{1}{m-n} \operatorname*{Min}_{x_k=c, x_j \geq 0} \left\{ \sum_{i=1}^m \frac{1}{\sigma_i^2} \right.$$
$$\left. \times \left[a(r_i) - \sum_{k=1}^n b_k(r_i) x_k \right]^2 \right\}. \quad (2)$$

In contrast to analogue projections of the χ^2-surface in nonlinear problems, $\chi_{r,k}^2(c)$ is a piecewise quadratic function of c for which analytic expressions can be easily derived [18] as a by-product of the calculation of the optimal parameter values $x_{\min}$ (e.g., a few matrix operations with the normal equations).

In principle, $\chi_{r,k}^2(c)$ follows a χ^2 probability distribution [18] and confidence intervals for each solution component $x_{k,\min}$ can be calculated [18]. However, when fitting data obtained with the analytical ultracentrifuge, it is more advantageous to directly inspect the resolution of the analysis by tracing $\chi_{r,k}^2(c)$, because of the absence of reliable values for the standard deviations and the asymmetry of $\chi_{r,k}^2(c)$ due to non-negativity of parameters. If the analysis is ill-conditioned, $\chi_{r,k}^2(c)$ will have a small ascent, indicating almost equal probability of all values of x_k in the neighborhood of $x_{k,\min}$; if the resolution is high, $\chi_{r,k}^2(c)$ will exhibit a sharp minimum around $x_{k,\min}$.

Since $\chi_{r,k}^2(c)$ is piecewise quadratic, in the neighborhood of $x_{k,\min}$ (i.e., until one parameter x_j eventually has been driven to zero by the constraint in x_k),

$$\chi_{r,k}^2(c) = \chi_{r,k}^2(x_{k,\min}) + \frac{1}{2}(c - x_{k,\min})^2 \frac{d^2 \chi_{r,k}^2}{dc^2} \quad (3)$$

is valid, where the second derivative does not depend on the data, but only on the basis functions and on the experimental conditions. It can be used as well as singular value decomposition [17] to optimize the experimental setup (e.g., to select the rotor speed for an optimal detection of a certain component), to predict ill-conditioning and to calculate estimations for the confidence intervals before the data acquisition.

For many purposes (e.g., if the binding stoichiometry is more important than the binding strength), this linear approach is suitable. However, if a confidence interval of an association constant K is of interest, $\chi_r^2(K)$ should be traced (with all parameters x optimized while the mass act relation according to K is valid). Although for $\chi_r^2(K)$ no analytic expression can be derived, it can be calculated with a small amount of computer time on a grid of K-values (preferably on a logarithmic scale), if the mass act relation is used as an additional (nonlinear) equality constraint imposed on the linear least squares problem. The technique of Lagrange multipliers and a combinatorial approach to keep the parameters nonnegative allow an efficient calculation of $\chi_r^2(K)$.

2.2 Fitting the two-dimensional data surface

The analysis of the data set composed of radial and wavelength profiles needs a description of the radial and spectral dependence of the absorbance of each component:

$$a_k(r, \lambda) = \varepsilon_k(\lambda) c_k(r) d$$
$$= \varepsilon_k(\lambda) c_k(r_0) d e^{\frac{M_k(1-\bar{v}_k\rho)}{2RT} \omega^2(r^2-r_0^2)} \quad (4)$$

(with a_k: absorbance of component k; r: radius; r_0: arbitrarily reference radius; λ: wavelength; $\varepsilon_k(\lambda)$: extinction profile; d: thickness of centerpiece; M_k: relative molar mass; $\bar{v}_k$: partial specific volume; ρ: buffer density; ω: angular velocity of the rotor; T: rotor temperature) [19]. Complexes formed by association equilibria are described as distinct components and thermodynamically ideal sedimentation behavior is assumed. The absorption measured is the sum of all K components' absorptions and, therefore

$$\underset{\varepsilon_k(\lambda_l),c_k(r_0),b(\lambda_l)}{\text{Min}} \sum_{i,l} \Bigg[A(r_i,\lambda_l) - b(\lambda_l)$$
$$- \sum_{k=1}^{K} a_k(r_i,\lambda_l) \Bigg]^2 \qquad (5)$$

has to be solved, where $A(r_i,\lambda_l)$ is the experimental absorbance at radius r_i and wavelength λ_l and $b(\lambda_l)$ a wavelength dependent background. Free parameters to minimize the sum of squared residuals are the concentrations $c_k(r_0)$ of all components at the reference radius, the extinction values $\varepsilon_k(\lambda_l)$ of each component at wavelengths λ_l and, in general, the background values $b(\lambda_l)$. For simplicity, the statistical weights were chosen to be equal to the mean standard deviation for all points and were omitted in the equations. The sum in (5) has to be taken over all experimental data points.

This separable nonlinear least squares problem includes the difficulty that the number of parameters to be determined is very high (e.g., each $\varepsilon_k(\lambda)$ in the analysis of the data shown in Fig. 1 amounts to a total of 170 parameters at a wavelength resolution of 1 nm). Therefore, in the case of many components additional constraints based on similarity (e.g., proportionality) of the $\varepsilon_k(\lambda)$ of certain components have to be used. During the calculations these additional constraints can be dropped partially by adjusting the resolution of the $\varepsilon_k(\lambda)$ according to the information content of the data.

To explain this concept in more detail: A first step reduces the number of unknown parameters of the $\varepsilon_k(\lambda)$ for each component to the number of different chromophores in solution. At this stage different techniques to determine the background $b(\lambda_l)$ appear. So far, no use is made of the radial distribution and no hypochromism or environment effects of the chromophores in particular components are considered. In step 2 starting values for the $c_k(r_0)$ and the $\varepsilon_k(\lambda_l)$ are calculated, alternatively without any prior knowledge (step 2a) or with use of prior knowledge (step 2b) about the extinction profiles and molar masses to increase the number of components that can be resolved. The iterative improvement of both $c_k(r_0)$ and $\varepsilon_k(\lambda_l)$ follows in step 3. The knowledge about the radial distribution enables an individual refinement of some of the $\varepsilon_k(\lambda_l)$, taking, for example,

hypochromism into account. An analysis of the statistical accuracy of the further approximations for the $\varepsilon_k(\lambda_l)$ is used to avoid overinterpretation of the data.

Step 1: Reduction of the number of unknown parameters of the extinction profiles: To reduce the number of unknown parameters of the $\varepsilon_k(\lambda)$ it is useful to assume that the absorption of analogue chromophores (or subunits) bound in different components is equal. This approximation is exactly true if no hypochromism or environment effects occur during complex formation. In this case, all measured absorbance profiles are different linear combinations of the same chromophore spectra. For clarity, it will be assumed that only two different types of chromophores are present in the solution. This case can be extended naturally without any further complication. With this assumption, (5) can be simplified to

$$\underset{\varepsilon_1,\varepsilon_2,b,f_r,g_r}{\text{Min}} \sum_{r,l} (A_r(\lambda_l) - b(\lambda_l) - f_r\varepsilon_1(\lambda_l) - g_r\varepsilon_2(\lambda_l))^2 ,$$
$$(6)$$

where A_r is the experimental absorbance of the wavelength scan at radius r, and $\varepsilon_1(\lambda)$, $\varepsilon_2(\lambda)$ are two chromophore absorbance profiles with the factors f_r and g_r of the linear combinations. Unfortunately, for (6) no unique solution $(\varepsilon_1^*, \varepsilon_2^*, b^*)$ exists, because the data only define a two-parametric subspace in function space which can be described equivalently by an infinite number of sets $(\varepsilon_1^*, \varepsilon_2^*, b^*)$. Nevertheless, if the concentration gradients of the chromophores are different and not too small, an arbitrarily constrained Marquardt procedure (e.g., perpendicular in the $(\varepsilon_1^*, \varepsilon_2^*, b^*)$) produces a well defined solution with very good convergence properties. The lasting uncertainty of the "true" profiles $(\varepsilon_1, \varepsilon_2, b)$ then can be expressed as

$$\varepsilon_i(\lambda_l) = \alpha_i \varepsilon_1^*(\lambda_l) + \beta_i \varepsilon_2^*(\lambda_l), \quad i = 1, 2 \qquad (7.1)$$

$$b(\lambda_l) = b^*(\lambda_l) + \alpha_b \varepsilon_1^*(\lambda_l) + \beta_b \varepsilon_2^*(\lambda_l) \qquad (7.2)$$

Of course, absorbance profiles scanned in separate experiments can also be used to serve as starting point for $\varepsilon_1(\lambda)$ and $\varepsilon_2(\lambda)$ in the iteration. However, it is advantageous – in particular with respect to the background absorption present in both experiments – to determine the chromophore absorbance profiles with a concentration gradient, and that the absorbance profiles are consistent with the experimental data to be evaluated.

The background absorption: The constants α_b and β_b in (7.2) are analogous to the baseline problem in the fit of a single radial profile. The background can be determined by computing α_a and β_b, which complicates the problem to the same extent as one additional component. For simplicity, these calculations can be performed by formally including one additional (pseudo-) component with relative molar mass zero and with an extinction profile equal to the righthand side of (7.2). This profile can be modified during the iteration similar to the extinction profiles of the components.

Unfortunately, this method worsens the resolution of the components under interest and it is advantageous, as in the analysis of a single radial profile, to use experimental techniques (e.g., the meniscus depletion technique or overspeeding at the end of the experiment) to determine $b(\lambda_l)$. For clarity, we omit the background problem in the following equations.

Step 2a: Calculation of starting values without a priori knowledge: With the use of (7.1), one only has to solve the linear problem

$$\underset{\alpha_k^0, \beta_k^0}{\text{Min}} \sum_{i, l} \Bigg(A(r_i, \lambda_l) - \sum_k d \left[\alpha_k^0 \varepsilon_1^*(\lambda_l) + \beta_k^0 \varepsilon_2^*(\lambda_l) \right]$$

$$\times e^{\frac{M_k(1 - \bar{v}_k\rho)}{2RT} \omega^2 (r^2 - r_0^2)} \Bigg), \tag{8}$$

where $c_k(r_0)$ and α_k, β_k are subsumed to new parameters α_k^0, β_k^0 respectively. In general, it is well-conditioned as long as the number of components and the distribution of molar masses follow the common restrictions for the analysis of a single sedimentation profile ($K \leq 3$–4, this can be quantified more exactly by the eigenvalue method of Twomey [20] or by singular value decomposition [17]). The result of the analysis are absorbance profiles corresponding to each component. They can be refined in further approximations as described in step 3. In simple cases (i.e., few components or not completely overlapping chromophores) even the molar masses can be estimated by applying nonlinear procedures (e.g., [21, 22]) to the radial scans at different wavelengths or, in some cases more advantageous [23], immediately

to (8). In these cases the two-dimensional data surface can be unravelled in components' contributions that are characterized relative to their molar masses and their absorption spectra, without any *a priori* knowledge. An example is given in Fig. 2.

Step 2b: Calculation of starting values with a priori knowledge about the extinction profiles and relative molar masses: If little *a priori* knowledge about the components' molar masses and absorption spectra is available, a more detailed analysis is possible. The unknown constants α_1, α_2, β_1 and β_2 in (7.1) can be determined by two boundary conditions for each chromophore. One of these conditions can be a normalization to a known literature extinction coefficient at a certain wavelength. Another one can be any characteristic information about the spectra, e.g., to be zero above 320 nm for proteins without absorbing prosthetic groups, the similarity to a absorbance profile measured in a separate experiment or the similarity to a profile at the meniscus (if only one chromophore is left at the meniscus). Subsequently, the $\varepsilon_k(\lambda_l)$ of all components can be calculated according to the stoichiometry of their chromophore (or subunit) content. Then, the problem (5) becomes linear:

$$\underset{c_k(r_0)}{\text{Min}} \sum_{i, l} \left[A(r_i, \lambda_l) - \sum_{k=1}^{K} a_k(r_i, \lambda_l) \right]^2, \tag{9}$$

and the solution will give first approximations for the $c_k(r_0)$. Again, singular value analysis can be used to estimate the number of components that can be resolved. Now, the problem is well-conditioned if the number of components *containing one kind of chromophore* follows the common restriction.

Step 3: Improvement of the concentrations and extinction profiles: Approximate knowledge about the components radial distribution can improve the calculation of the extinction profiles, considering hypochromism effects. With given $c_k(r)$, (5) splits up into separate linear problems for each wavelength grid:

$$\underset{\varepsilon_k(\lambda_l)}{\text{Min}} \sum_{i, l} \left(A(r_i, \lambda_l) - \sum_k dc_k(r_i) \varepsilon_k(\lambda_l) \right)^2$$

$$= \sum_l \underset{\varepsilon_k(\lambda_l)}{\text{Min}} \sum_i \left(A(r_i, \lambda_l) - \sum_k dc_k(r_i) \varepsilon_k(\lambda_l) \right)^2. \tag{10}$$

With

$$\Delta\varepsilon_k(\lambda) = \varepsilon_k^{(1)}(\lambda) - \varepsilon_k^{(0)}(\lambda) , \tag{11}$$

$$\min_{\Delta\varepsilon_k(\lambda_l)} \sum_i \left(\Delta A(r_i, \lambda_l) - \sum_k dc_k(r_i)\,\Delta\varepsilon_k(\lambda_l) \right)^2$$

$$\text{for all } l \tag{12}$$

have to be solved, where $\Delta A(r_i, \lambda_l)$ is the deviation of experimental and calculated data in the previous approximation, and $\Delta\varepsilon_k(\lambda_l)$ is the difference of $\varepsilon_k(\lambda_l)$ in the new and the previous approximation. Unfortunately, (12) can – analogously to (8) – only be solved if all molar masses are different (otherwise the basis functions $dc_k(r)$ of (12) will be linearly dependent). Additional problems may arise if the concentrations of some components are low (or even zero), indicating the lack of information about their absorption spectra within the data. For these reasons, the restrictions used in the first approximation (no hypochromism) can be dropped only partially. But often suitable assumptions can be made about components that will not show hypochromism effects, or groups of components that have the same subunit spectra. This leads to new correlated corrections $\Delta\varepsilon_n(\lambda)$ which are linked by the stoichiometry $s_{k,n}$ to the corrections for the components spectra $\Delta\varepsilon_k(\lambda)$:

$$\Delta\varepsilon_k(\lambda) = \sum_n s_{k,n}\,\Delta\varepsilon_n(\lambda) , \tag{13}$$

and which modify the structure of the problems (12):

$$\min_{\Delta\varepsilon_n(\lambda_l)} \sum_i \left(\Delta A(r_i, \lambda_l) \right.$$

$$\left. - d \sum_n \left\{ \sum_k c_k(r_i) s_{k,n} \right\} \Delta\varepsilon_n(\lambda_l) \right)^2$$

$$\text{for all } l . \tag{14}$$

The possibility to compute more detailed extinction profiles can be quantified by applying (3) to the problems (14). As the second derivative in (3) applied on (14) is independent of the wavelength, the conditioning of (14) will be equal for all wavelength grids, and confidence intervals for the $\Delta\varepsilon_n(\lambda)$ can be calculated. At this point it is useful to investigate the information content

for all reasonable correlations and make a compromise between a detailed analysis of the $\varepsilon_k(\lambda)$ and their stability with respect to the experimental error.

The refinement of the $\varepsilon_k(\lambda)$ is followed by the computation of more accurate $c_k(r_0)$ via (9) and corresponding new χ_r^2. This iteration step can be combined with a line test with respect to a decrease in the χ_r^2, and the step-size can be adjusted, e.g.

$$\varepsilon_k^{(1)} = \varepsilon_k^{(0)} + \left(\frac{1}{2}\right)^n \Delta\varepsilon_k . \tag{15}$$

This iteration step can be repeated until no further improvement in χ_r^2 is achieved. The convergence of the iteration process is rapid (~ 10 steps) and will depend on the degree of hypochromism or environment effects that the components exhibit. Finally, an analysis of the statistical accuracy of the $c_k(r_0)$ by tracing $\chi_{r,k}^2(c)$ shows the information content on each component and the degree of exchangeability of the contribution to the fit, inclusive the spectral exchangeability.

The entire procedure, with data consisting of 1000 data points and running on a SCHEME interpreter installed on a 386 PC 40 MHz, takes about 2–3 h.

Limits of the method: Limits of this method are obvious if some of the components' $a(r, \lambda)$ are linearly dependent, e.g.,

$$a_k(r, \lambda) = \alpha_1 a_m(r, \lambda) + \alpha_2 a_n(r, \lambda) , \tag{16}$$

at least in the order of the noise of the data acquisition. This requires that

$$M_k \approx M_m \approx M_n \tag{17.1}$$

$$\varepsilon_k(\lambda) \approx \alpha_1 \varepsilon_m(\lambda) + \alpha_2 \varepsilon_n(\lambda) \tag{17.2}$$

are simultaneously fulfilled, where the possibility to distinguish molar masses and absorbance spectra additionally depends on the components' concentrations. For example, protein-ligand interactions with relative small ligand molar mass can comply with (17.1) and (17.2), if complexes with different ligand content have to be considered. As (17.2) defines a two-dimensional subspace in function space, it can be represented in the analysis by two independent components. The sum of the contributions of these two components to the fit then equals an average contribution of the entire subspace to the fit (e.g., an average ligand occupation) and can be subject to further interpretations.

3. Experimental

3.1 Material

The band 3 protein of the human erythrocyte membrane was solubilized and purified in solutions of $C_{12}E_9$ as described in detail in [13]. In this non-ionic detergent, the band 3 protein is in a monomer/dimer/tetramer association equilibrium for approximately 2 days [13]. Buffer composition was 10 mM NaCl, 5 mM phosphate buffer (pH 8.0), 0.4% $C_{12}E_9$. DBDS was prepared according to [24] and was a kind gift of Prof. H. Fasold, Institut für Biochemie, J.W. Goethe-Universität Frankfurt/Main. HbO was prepared as described in [9].

3.2 Analytical ultracentrifugation

Sedimentation equilibrium experiments were performed in a Beckman Optima XL-A using 12 mm 6-channel centerpieces. Sample volume was 130 μl, rotor speed 12 500 (resp. 11 000) rpm, rotor temperature 4 °C. Data acquisition took 1/2–1 h per experiment. Control experiments with band 3 showed that, for the preparation used, the monomer/dimer association constant was in the order of 0.5 μM and the dimer/tetramer association constant in the order of 5 μM. For total loading concentrations of 2.5 μM, a sample volume of 130 μM and a rotor speed of 12 500 rpm, the mean concentrations of all band 3 oligomers in the sample were approx. 30% of the total loading concentration. Control experiments with DBDS showed that in 5 mM phosphate buffer (pH 8.0), 10 mM NaCl, 0.4% $C_{12}E_9$, DBDS was partially bound to detergent micelles which sedimented with reduced molar mass $M_r^* = 13\,000$ (for simplicity relative molar masses M_r are transformed to reduced molar masses M_r^* according to $M_r(1 - \bar{v}\rho) = M_r^*(1 - \bar{v}^*\rho)$ with hypothetical assumed partial specific volume $\bar{v}^*$ of 0.73 cm^3 g^{-1} [1]).

3.3 Validity of Lambert–Beer's law

Nonlinearities in the dependence of the measured absorption on the concentration of the solutes in general depend on the wavelength used for measurement and on the steepness of the extinction profile of the solutes. Therefore, nonlinearities will produce concentration dependent deformations of the measured absorption spectra. This is particularly true for the effect of finite bandwidth of the monochromator (3 − 4.5 nm).

If one produces a concentration gradient of a chromophore in the analytical ultracentrifuge and compares the absorption profiles at several radii, it is possible to test the validity of Lambert–Beer's law without the use of an external photometer. Provided that the measured absorption spectra are proportional within the detection limits, Lambert–Beer's law is valid. If broadening of the peaks at higher concentrations is visible, nonlinearities occur. Several absorption bands of hemoglobin in the visible range and in the UV and the UV absorption of the band 3 protein were used. Linearity of the absorption with respect to the concentration was observed at least up to 1.7 OD in the UV and 1.8 OD in the visible wavelength range.

4. Results

4.1 Experiments on samples containing two chromophores

Figure 1 shows the two-dimensional data surface of a mixture of the band 3 protein with a molar excess of DBDS in sedimentation equilibrium. To analyze the data without any *a priori* knowledge, a standard nonlinear double-exponential procedure was applied on the radial scans at 280 nm and 270 nm, yielding reduced molar masses M_r^* of 6980 and 213700. The iterative evaluation according to (6), (8) and (12) converged with a goodness of fit parameter χ_r^2 of 2.15, indicating the two-species model to give a quite realistic picture of the solutes. An integration of the absorption from the meniscus to the bottom with assumed mass conservation followed the computation of the $a_k(r_0, \lambda)$ to transform the absorption distributions into average optical densities $\bar{a}_k(\lambda)$. The calculated absorption spectra

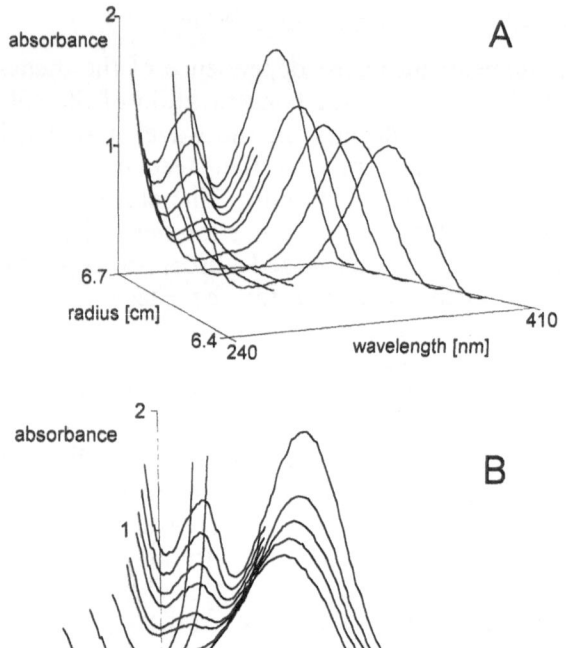

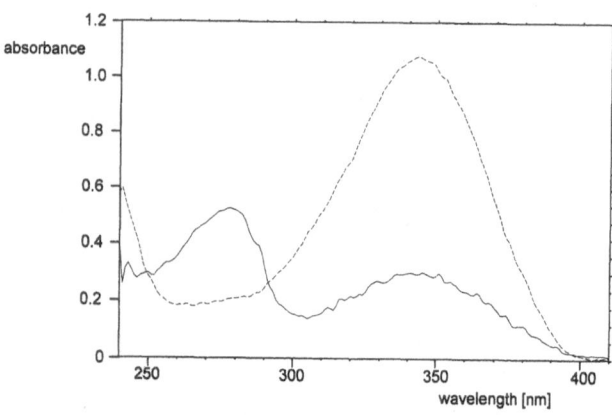

Fig. 1A). Experimental radial and wavelength dependent absorption of a mixture of band 3 protein (initial concentration $c_0 = 4.3\ \mu M$) and DBDS ($c_0 = 21.7\ \mu M$) in 10 mM NaCl, 5 mM phosphate buffer (pH 8.0), 0.4% $C_{12}E_9$, in sedimentation equilibrium at 12 500 rpm, 4 °C. B) same data viewed from a different perspective.

Fig. 2. Decomposition of the data shown in Fig. 1 without *a priori* knowledge into components with reduced molar mass M_r^* 213700 (——) and 6980 (---). The lines indicate the average absorption $\bar{a}_k(\lambda)$ in the cell, as obtained by integration over the cell volume. These components fit the data with $\chi_r^2 = 2.15$

(Fig. 2) show that part of the DBDS (with the typical 340 nm absorption band) is bound to the band 3 protein and sedimented with the higher molar mass.

A much more detailed analysis of the same data is possible, if the knowledge about the reduced molar masses of all possible components of the band 3/DBDS-interaction is used and the ambiguity (7.1) of the spectra is removed. Protein spectra were normalized to $\varepsilon_{280} = 81\ mM^{-1}\ cm^{-1}$ (based on [25]) and set zero above 320 nm. Since no protein containing components were left at the meniscus, it was possible to specify the DBDS spectrum by a least squares fit to the spectrum near the meniscus (using α_1 and β_1 of (7.1) as free parameters), and by normalizing to $\varepsilon_{336} = 50\ mM^{-1}\ cm^{-1}$ [26]. Table 1 shows all components with according reduced molar masses M_r^*. Because band 3/DBDS complexes with different stoichiometry satisfy conditions (17.1) and (17.2) for linearly dependent $a(r, \lambda)$, only unliganded band 3 and band 3/DBDS complexes with 1:1 stoichiometry were considered in the evaluation (this assumption is technically useful and without influence on the result, see above).

The calculated average concentrations in the cell (using (9), followed by averaging over the cell volume, see Table 1) indicate that the band 3 association equilibrium is shifted to the dimer, which apparently is the nearly exclusive binding site of DBDS. As the concentration of unliganded band 3 dimer is zero, it can be concluded that two DBDS molecules are bound at one band 3 dimer. For details see [27; Schuck P, Legrum B, Schubert D, Passow H, in preparation].

A comparison of the two rows of Table 1 shows the relative stability of the calculated concentrations with respect to minor changes in the $\varepsilon_k(\lambda_l)$ during the iteration. Because of the similar concentration gradient, the extinction profiles of DBDS in solution and in detergent micelles could not be resolved. If these profiles would be distinguished in the analysis, they would be unstable to noise. The calculated standard deviation for the extinction profiles (using (3)) would have been $30\ mM^{-1}\ cm^{-1}$ for free DBDS, $4.6\ mM^{-1}\ cm^{-1}$ for DBDS in micelles and $2.8\ mM^{-1}\ cm^{-1}$ for DBDS bound to band 3 protein. Therefore, (13) was applied with a $\Delta\varepsilon_1$ for DBDS in solution or bound in detergent micelles and a $\Delta\varepsilon_2$ for all band 3-bound DBDS, whereas the protein extinction

Table 1. Results of a fit to the data shown in Fig. 1 using knowledge about extinction profiles and relative molar masses of possible components of the band 3/DBDS-interaction.

Iteration step	χ_r^2	Average concentration [μM]								
		Component	DBDS in solution	DBDS in detergent micelle	Band 3 monomer	Band 3 dimer	Band 3 tetramer	band 3 monomer + 1 DBDS	Band 3 dimer + 2 DBDS	Band 3 tetramer + 4 DBDS
		M_r^*	100	13000	108000	216000	432000	108100	216200	432400
1	2.53		2.2	15.3	0	0.07	0.1	0	2.1	0
7	1.43		2.3	15.1	0	0	0.02	0	2.1	0.06

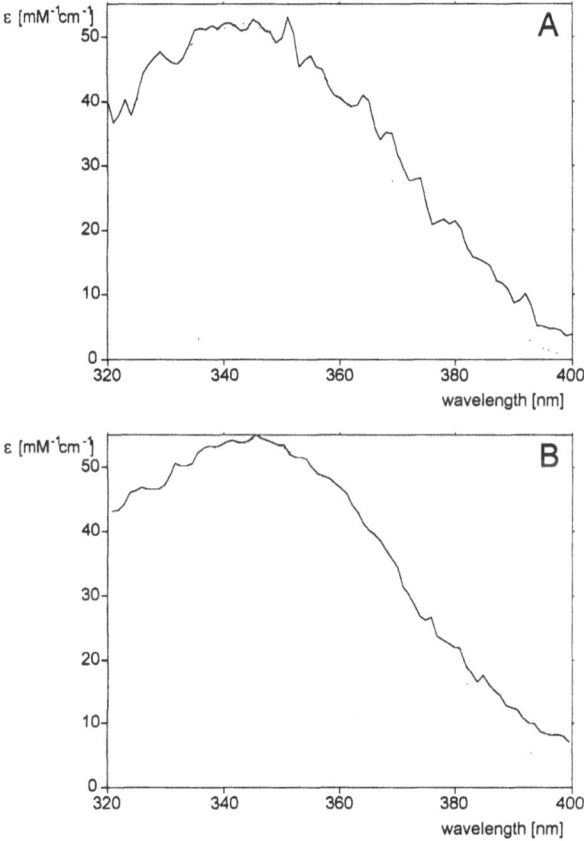

Fig. 3. Two pairs of calculated extinction profiles of DBDS in solution or detergent micelles ($\cdots$) or bound to the band 3 dimer (——), normalized to the same peak height. A) Analysis of the data shown in Fig. 1 B) Evaluation of an experiment with initial concentration 5.2 μM DBDS and 3.6 μM band 3 protein (in 10 mM NaCl, 5 mM phosphate buffer (pH 8.0), 0.4% $C_{12}E_9$; 12500 rpm, 4 °C); data not shown. To reduce the standard deviation of the calculated extinction profiles, a higher number of absorption profiles (17) was used. The absorbance at the bottom of the cell at 340 nm was 0.49.

was considered as unchanged during DBDS binding. The predicted standard deviation of the extinction profiles then was 0.35 mM^{-1} cm^{-1} for the unbound and 1.6 mM^{-1} cm^{-1} for the bound DBDS, indicating that spectra changes of protein bound DBDS can be resolved. This improved the quality of the fit from initially $\chi_r^2 = 2.53$ after 7 steps to $\chi_r^2 = 1.43$. As shown in Fig. 3A, the extinction profile of DBDS (similar to a monobenzoylated derivative [24]) apparently depends on the local environment and can be modified by binding to band 3. Artefacts due to nonlinearities in the optical system were excluded by experiments with lower DBDS concentration (Fig. 3B). To reduce the standard deviation of the calculated extinction profiles, in these experiments an enhanced number of wavelength profiles were included in the data set.

The residuals of the fit are shown in Fig. 4. Corresponding to the final χ_r^2 value near unity, they are in the range of the uncertainty of the data acquisition and evenly distributed. Enhanced noise is visible in the flanks of the spectra and in the steep parts of the radial scans.

The statistical accuracy of the results is illustrated by tracing, according to (2), χ_r^2 as a function of a particular constraint value c, whereas all other components were optimized (Fig. 5A). Although the distribution of DBDS in solution and into micelles could not be fully resolved, the other components all have a clearly defined minimum and are satisfactorily resolved, indicating a well-conditioned analysis. It should be noted that although some components have concentration zero, they nevertheless have to be resolved. They can worsen the conditioning of the analysis due to a possible exchangeability of their $a_k(r, \lambda)$. An

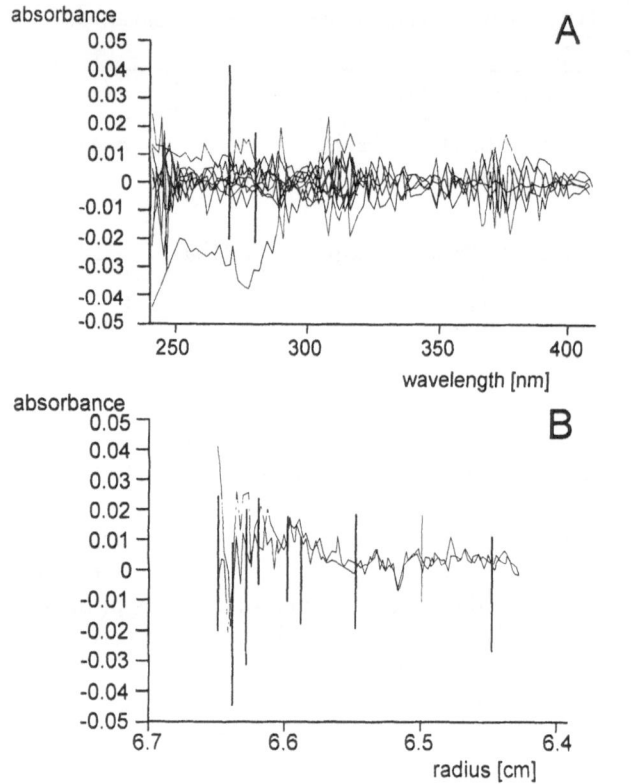

Fig. 4. Residuals of the fit to the data in Fig. 1 (results shown in Table 1, $\chi_r^2 = 1.43$). A) Projection into the absorbance-wavelength plane, B) Projection into the absorbance-radius plane

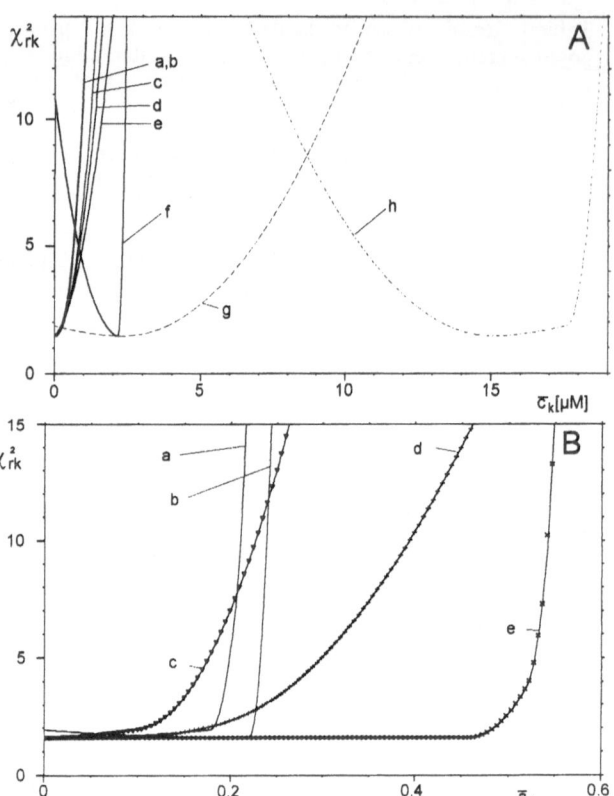

Fig. 5. Traces of the goodness of fit parameter $\chi_{r,k}^2(c)$ according to (2) for components listed in Table 1. The fit parameters $c_k(r_0)$ are finally transformed to average concentrations in the cell $\bar{c}_k$. A) Fit to the two-dimensional data surface shown in Fig. 1: a) band 3 monomer, b) band 3 monomer/1DBDS, c) band 3 dimer, d) band 3 tetramer, e) band 3 tetramer/4DBDS f) band 3 dimer/2DBDS, g) DBDS in solution, h) DBDS in detergent micelles. B) Fit exclusively to the radial profile at 280 nm (fit parameters are transformed to average absorbances in the cell $\bar{a}_k$). The curves for liganded (marks) and unliganded band 3 oligomers (solid lines) completely overlap: a) DBDS in solution, b) DBDS in detergent micelles, c) (liganded) band 3 monomer, d) (liganded) band 3 dimer, e) (liganded) band 3 tetramer

analogue representation of a fit to the single radial sedimentation profile at 280 nm (Fig. 5B) depicts the lack of information in this data set: no component has a satisfactory minimum, no mixed components can be resolved.

4.2 Experiments on samples containing three chromophores

Although allosteric interactions between band 3-bound DBDS and HbO binding have been supposed [27], the intention of the analysis of mixtures of band 3, DBDS and HbO here was mainly to investigate the resolution of sedimentation equilibrium experiments in the case of a superposition of a third chromophore.

To evaluate sedimentation equilibrium data of mixtures of band 3 protein, DBDS and HbO (Fig. 6), one has to consider 1) the interaction of the band 3 protein with DBDS (the components

are: band 3 monomer, band 3 dimer and band 3 tetramer, DBDS in solution, in detergent micelles and bound to the band 3 dimer, as analyzed above), 2) the interaction of the band 3 protein with HbO [9] (additionally HbO dimer, HbO tetramer and a band 3 tetramer/HbO dimer complex), and 3) a possible formation of mixed complexes with band 3 dimer/2DBDS/HbO dimer and band 3 tetramer/HbO dimer/2DBDS. Mixed complexes of all three species with other DBDS or HbO content again have nearly linearly

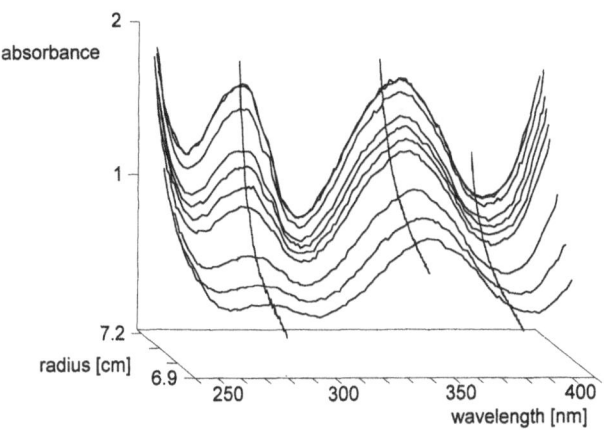

Fig. 6. Sedimentation equilibrium data of a mixture of 3.5 μM band 3 protein, 13.8 μM DBDS and 7.7 μM (heme) oxyhemoglobin in 10 mM NaCl, 5 mM phosphate buffer (pH 8.0), 0.4% $C_{12}E_9$. Rotor speed was 11 000 rpm, rotor temperature 4 °C. For clarity, projections of the radial profiles into the radius-wavelength plane (·····) were included in the graphics

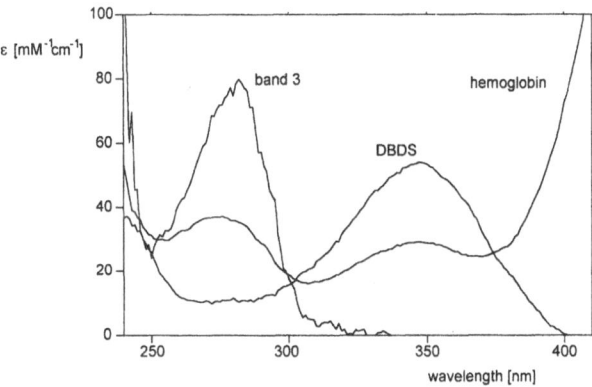

Fig. 7. Calculated extinction profiles of band 3 protein, DBDS and HbO, based on the analysis of the data shown in Fig. 6

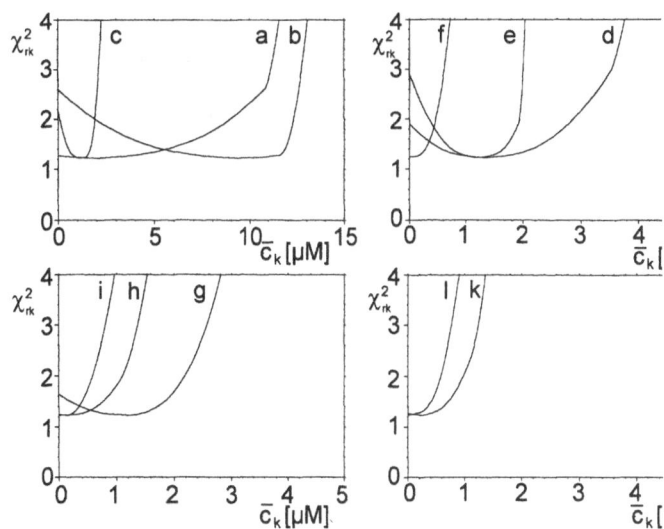

Fig. 8. Traces of $\chi^2_{r,k}(c)$ for each parameter of the fit to the data shown in Fig. 6 (the $c_k(r_0)$ are finally transformed to average concentrations $\bar{c}_k$ in the cell). a) DBDS in solution, b) DBDS in detergent micelles, c) band 3 dimer/2DBDS complex, d) HbO dimer, e) HbO tetramer, f) band 3 tetramer/HbO dimer complex, g) band 3 monomer, h) band 3 dimer, i) band 3 tetramer, k) band 3 dimer/2DBDS/HbO dimer complex, l) band 3 tetramer/HbO dimer/2DBDS complex.

dependent $a_k(r, \lambda)$ and, therefore, their possible contribution to the fit is represented by the components mentioned before.

Starting values for the $\varepsilon_k(\lambda_l)$ in this case were calculated with data of separate experiments, each one according to (6). The final extinction profiles (after convergence was achieved with a χ^2_r of 1.22) are shown in Fig. 7, the results and their statistical accuracy in Fig. 8. It is obvious that part of band 3 is liganded with DBDS and the unliganded band 3 is in the monomeric state. HbO is in a dimer/tetramer equilibrium and, at the HbO concentrations used in the experiment, no significant

amount of HbO-containing mixed complexes could be detected. All curves $\chi^2_{r,k}(c)$ exhibit a clear minimum (except the two curves describing DBDS in solution and in detergent micelles, similar to Fig. 5a), indicating independent resolution of all 11 components. Nevertheless, compared with the analysis of the band3-DBDS interaction with only two chromophores, the superposition with the HbO-band 3 association slightly reduces the resolution. In particular, the differences in the extinction profiles of band 3-bound DBDS and DBDS in solution could not be resolved.

Discussion

We have described a method to analyze sedimentation equilibrium data consisting of several absorption profiles measured at various radii and several radial profiles measured at various wavelengths. In sedimentation equilibrium these data can be conveniently acquired. The results clearly indicate the usefulness of this novel extension of the data space for the investigation of species with different absorption spectra. The resolution is greatly enhanced, because the possibility to distinguish components by their absorption spectra can stabilize an otherwise ill-conditioned analysis.

The potential of the method was demonstrated by studies on the band 3-DBDS interaction. In this system, only small differences in the molar masses of liganded and unliganded protein are present, nevertheless, all components can be resolved. However, these small differences are not a prerequisite for the method but show its resolution power. In this special application even an analysis of a single radial profile at 340 nm would have identified the band 3 dimer as the exclusive binding site for DBDS. The maximum stoichiometry of 2 DBDS/band 3 dimer and the spectral changes of bound DBDS due to interaction with band 3 protein would, however, not have been unravelled. On the other hand, the 11 components, which have to be considered in the analysis after addition of hemoglobin, could not have been resolved following standard procedures, whereas the procedure described in the present paper can solve the problem.

The resolution of components depends on the amount of *a priori* knowledge that can be introduced into the calculations. It can be adjusted to the specific conditions, changing continually between an analysis without any prior knowledge, which resolves a few components, yielding both (reduced) molar masses and absorption spectra, and an analysis which resolves a number of components that is up to 3–4 times the number of different chromophores in the solution, if relative molar masses, partial specific volumes, and a little characteristic information about the absorption spectra is available. In addition to the determination of concentrations, complete absorption spectra of all components are the result of an analysis, partially including the detection of hypochromism or environment effects of the chromophores.

It is clear that even the present method has its limitations: Not resolvable are components that have similar molar masses and have linearly dependent extinction profiles. Furthermore, the information included in different absorption profiles is redundant for species with similar extinction profiles, and the resolution is not enhanced. Therefore, if only the radial distribution is of interest, in principle, several radial profiles

measured at different characteristic wavelengths contain the same information as the two-dimensional data surface. The evaluation of such data sets can be performed according to [10–12]. This requires *a priori* knowledge of at least two (dependent on the number of radial scans used in the analysis) extinction coefficients for each component, which have to be determined in separate experiments. Unfortunately, the result will be sensitive to their values (particularly in experiments with more than two chromophores).

The use of absorption profiles offers two advantages: Firstly, to distinguish the components by their absorption profiles, the use of wavelength scans automatically generates a large statistical base of extinction data. Compared to only two separately measured extinction coefficients the statistic is included in the original data and no additional experiments are necessary. Only one absolute extinction value is needed to normalize the results. Secondly, the exclusive use of radial profiles fails if, for example, hypochromism changes the absorption profiles of some components. A possibility to circumvent this difficulty could be to scan the radial profiles at previously known isosbestic points [11]. This may be difficult on the flanks of the absorption profiles (see the increased noise indicated by the data shown) or if the number of radial scans necessary is higher than the number of isosbestic points. In contrast, the results show that the use of absorption profiles stabilizes the analysis against (at least small) changes in absorption spectra.

In principle, the use of a two-dimensional data surface as described in the present paper is open to be incorporated into many other methods of data analysis in analytical ultracentrifugation, since the evaluation does not depend on a specific radial distribution.

In studies on continuous molar mass distributions of synthetic copolymers it can be expected that polymers with equal molar mass distribution but with different absorption spectra can be resolved, similarly to the discrete case shown above. The described procedure can be modified by replacing the single Boltzmann term for the radial distribution of a component in (4) by a suitable integral over the molar mass distribution of a polymer. The well-known difficulty of the instability in the evaluation of the radial profiles then appears – separate from the evaluation of the absorption profiles and partially stabilized by the spectral information – in (8) and (9), making techniques as regularization [28] or additional models of the molar mass distributions [29] necessary. Similar modifications in the description of the radial distribution in (4) enable the application to non-ideal solutions.

For many purposes, the simultaneous analysis of radial profiles of multiple experiments with different rotor speeds and initial concentrations exhibits the optimal way of using information on the radial distribution [2, 23]. It is a particularly powerful method in the determination of interaction properties of biological macromolecules. A very promising approach would be the simultaneous analysis of two-dimensional data surfaces of multiple experiments, a natural extension of the simultaneous analysis of multiple radial profiles. In addition (and in a way orthogonal) to the well-known advantages, it should be possible to attain a spectral resolution of components. This would considerably increase the number of components that can be characterized simultaneously. The use of the multiple spectral information contained in the data surfaces would require some extensions in the calculations of the extinction profiles in (6) and (10)–(14), where the background absorptions and the component's concentrations are local to each data surface and the extinction profiles and the association constants are global parameters.

The method described is advantageous only in the investigation of heterogeneous interactions of species which possess different absorption spectra, e.g., many ligand/protein, protein/protein, and DNA/protein interactions. Labelling of species [2, 6–8] with conservation of interaction properties will, however, extend the range of useful applications.

Acknowledgements

The author is grateful to Prof. D. Schubert, Prof. J. Baumeister, and Dr. M. Zulauf for stimulating discussions, and to the Studienstiftung des deutschen Volkes, the Deutsche Forschungsgemeinschaft (SFB 169), and Beckman Instruments for financial suport. And he acknowledges with pleasure the suggestions of Dr. V. Burwitz concerning the manuscript.

References

1. Schubert D, Schuck P (1991) Colloid Polym Sci 86:12–22
2. Hsu CS, Minton AP (1991) J Mol Recogn 4:93–104
3. Varah JM (1985) SIAM J Sci Stat Comput 6:30–44
4. Hanlon S, Lamers K, Lauterbach G, Johnson R, Schachman HK (1962) Arch Biochem Biophys 99:157–174
5. Schachman HK, Gropper L, Hanlon S, Putney F (1962) Arch Biochem Biophys 99:175–190
6. Osborne JC, Powell GM, Brewer HB (1980) Biochim Biophys Acta 619:559–571
7. Mulzer K, Kampmann L, Petrasch P, Schubert D (1990) Colloid Polym Sci 268:60–64
8. Laue TM, Senear DF, Eaton S, Ross JBA (1993) Biochemistry 32:2469–2472
9. Schuck P, Schubert D (1991) FEBS Lett 293:81–84
10. Servillo L, Brewer HB, Osborne JC (1981) Biophys Chem 13:29–38
11. Lewis MS (1993) Abstracts, VIII International Symposium on Ultracentrifugation, Osnabrück
12. Lakatos S, Minton P (1991) J Biol Chem 266:18707–18713
13. Pappert G, Schubert D (1983) Biochim Biophys Acta 730:32–40
14. Passow H (1986) Rev Physiol Biochem Pharmacol 103:61–203
15. Johnson ML, Correia JJ, Yphantis DA, Halvorson HR (1981) Biophys J 36:575–588
16. Johnson ML, Lindsay MF (1992) Methods Enzymol 210:1–37
17. Lawson CL, Hanson RJ (1974) Solving Least Squares Problems. Prentice-Hall, Englewood Cliffs, New Jersey
18. O'Leary DP, Rust BW (1986) SIAM J Sci Stat Comput 7:473–489
19. Fujita H (1975) Foundations of Utracentrifugal Analysis. John Wiley & Sons, New York
20. Twomey S, Howell HB (1967) Appl Opt 6:2125–2131
21. Evans JW, Gragg WB, LeVeque RJ (1980) Math Comp 34:203–211
22. Ruhe A (1980) SIAM J Sci Stat Comput 1:481–498
23. Beechem JM (1992) Methods Enzymol 210:37–67
24. Kotaki A, Naoi M, Yagi K (1971) Biochim Biophys Acta 229:547–556
25. Dorst HJ, Schubert D (1979) Hoppe-Seyler's Z Physiol Chem 360:1605–1618
26. Casey JR, Lieberman DM, Reithmeier AF (1989) Methods Enzymol 173:494–512
27. Salhany JM (1990) Erythrocyte band 3 protein. CRC Press, Boca Raton, Florida
28. Provencher SW (1982) Comput Phys Commun 27:213–227
29. Lechner MD, Mächtle W (1991) Progr Colloid Polym Sci 86:62–69

Received June 26, 1993
accepted October 25, 1993

Authors' address:

Peter Schuck
Institut für Biophysik der
J.W. Goethe-Universität Frankfurt
Theodor-Stern-Kai 7, Haus 74
60590 Frankfurt/Main, Germany

Progress in Colloid & Polymer Science

Progr Colloid Polym Sci 94:14–19 (1994)

Determination of the molar mass of pigment-containing complexes of intrinsic membrane proteins: Problems, solutions and application to the light-harvesting complex B800/820 of Rhodospirillum molischianum

D. Schubert[1,2]), C. Tziatzios[1]), J.A. van den Broek[1]), P. Schuck[1]), L. Germeroth[2]) and H. Michel[2])

Institut für Biophysik, JWG-Universität[1]), and MPI für Biophysik[2]), Frankfurt am Main, FRG

Abstract: An adaptation of a procedure for the simultaneous determination of partial specific volumes and molar masses by sedimentation equilibrium experiments (Edelstein SJ, Schachman HK (1967) J Biol Chem 242: 306–311) to intrinsic membrane proteins in detergent solutions is described. The method requires (1) that the protein is available in at least two detergents the densities of which differ as much as possible, and (2) that the effective molar masses of the mixed protein/detergent micelles, $M(1 - \bar{v}\rho_0)$, are determined under conditions of density matching. The method is applied to a pigment-containing membrane protein, the light-harvesting complex B800/820 (LH III) from Rhodospirillum molischianum, in solutions of three detergents. It is shown that the complex is the hexamer of the basic α/β/pigment unit.

Key words: Sedimentation equilibrium – partial specific volume – molar mass – protein/pigment/detergent micelles – light-harvesting complex B800/820

Introduction

Pigment-containing membrane proteins, in particular the photosynthetic reaction centers of plants and certain bacteria and their associated light-harvesting complexes, represent in most cases stable aggregates of two to four different polypeptide chains which, in turn, may form well-defined stable oligomers [1, 2]. Undoubtedly, the determination of the molar mass of the particles is a useful task. Since, as is now generally agreed, solubilization of membrane proteins by suitable nonionic detergents does not disturb their native secondary, tertiary and quaternary structure, the molar mass determination can be performed on the detergent-solubilized proteins [3, 4]. Sedimentation equilibrium analysis in the analytical ultracentrifuge appears to be the most suitable method for that purpose [4, 5].

When pigment-containing membrane proteins, in detergent solution, are studied by sedimentation equilibrium analysis, one has to take into account: (1) The contribution of the membrane-bound detergent to the relative molar mass M and to the partial specific volume $\bar{v}$ of the protein/pigment complex; (2) the contribution of the pigment to $\bar{v}$. To solve the former, solutions have been devised by Tanford and Reynolds [4]. We will describe how detergent densities, crucial in this method, can be determined by using an analytical ultracentrifuge equipped only with absorption optics. We will then describe an adaptation of a well-known but rarely applied method of Edelstein and Schachman [6] to solve problem (2). Finally, the methodology outlined will be applied to the light-harvesting complex B800-820 (LH III) from Rhodospirillum molischianum.

Experimental

The detergents used: $C_{12}E_9$ (nonaethyleneglycol lauryl ether), LDAO (N,N-dimethyldodecylamine N-oxide) and $C_{12}M$ (dodecyl-ß-D-maltoside), were from Sigma (Munich), Fluka (Neu-Ulm) and Biomol (Hamburg), respectively.

D_2O (99.7%) was a gift of the Karlsruhe Nuclear Research Centre, and $D_2{}^{18}O$ (98% ^{18}O) was purchased from Johnson Matthey (Karlsruhe). The dye 1,6-diphenyl-1,3,5-hexatrien (DPH) was obtained from Serva (Heidelberg). All other reagents were from Merck (Darmstadt) and were of analytical grade.

The isolation and purification of LH III from Rhodosprillum molischianum strain DSM 120 was performed as described earlier [7, 8]. Exchange of the detergent was done during the final ion exchange chromatography. Detergent concentration in most cases was 0.1%. Protein concentration was determined from its absorbance at 370 nm ($^{1\%}_{1\,cm}$ A(370 nm) of the protein is approx. 200 [7, 8]) and was around 15 μg/ml. Buffer was 20 mM Tris-HCl (pH 8.5), 50 mM NaCl.

Sedimentation equilibrium experiments were performed in a Beckman Optima XL-A analytical ultracentrifuge, using 6-channel centerpieces. Sample volume was 100 μl, rotor speed between 15000 and 25000 rpm, and rotor temperature 6 °C. The absorbance-versus-radius data were recorded at 357 nm (DPH) or 370 nm (LH III) and were evaluated as described [9]. Solvent densities were determined by means of a Paar DMA 02 densitometer.

Determination of detergent densities

To establish the molar mass of the protein/pigment part of uniform protein/pigment/detergent complexes by sedimentation equilibrium analysis, either the experiments have to be performed at a solvent density which equals that of the detergent, or the effective molar mass of the complex, $M_{eff} = M(1 - \bar{v}\rho_0)$, has to be measured at different solvent densities ρ_0 and then to be extrapolated to the detergent density [4, 5]. Density adjustment is achieved by adding D_2O or $D_2{}^{18}O$ [4, 6]. Establishing the detergent density is a critical factor in this method. Here, sedimentation equilibrium analysis can be applied, by determining the solvent density at which the effective molar mass of the protein-free detergent micelles is zero [4, 5]. The use of interference optics has been advocated for that purposes [4]; however, currently available analytical ultracentrifuges are equipped with absorption optics only. They can

be directly applied to detergents absorbing in the easily accessible wavelength range, like Triton X-100 [10, 11]. Other detergents have to be stained in order to make the micelles detectable by the optical system. The "dye"/detergent ratio should be low enough not to change significantly the micelle density. Initially, we have used phenol to stain the micelles [5, 9]; however, in subsequent experiments reproducibility was found to be poor. We now consider DPH as a good choice (except with octylglucopyranoside and related compounds, because of poor solubility): (1) Its specific absorbance (at 357 nm) is high; (2) interference with absorbing impurities in commercial batches of D_2O and $D_2{}^{18}O$ ([6] and own observations) is negligible; (3) due to its hydrophobic nature, it can be expected to partition completely into the micelles, which minimizes baseline errors (but makes it necessary to prepare the stock detergent/dye mixture in an organic solvent). Results obtained with LDAO, $C_{12}E_9$ and $C_{12}M$ are shown in Fig. 1. The data for $C_{12}M$ have, however, to be corrected for H-D exchange (see below).

The partial specific volume of protein/pigment complexes in detergent solution

Most published partial specific volumes $\bar{v}$ of proteins were determined by direct density measurements [12, 13], in all recent studies by using the Paar densitometer [14]. With intrinsic membrane proteins in detergent solutions, however, this method seems to be of dubious value (see the obviously incorrect results described in [15]). According to our experience, the problems arising may result from detergent adsorption to the walls of the measuring cell. An alternative method is to calculate $\bar{v}$ according to the classical Cohn-and-Edsall procedure, applying Traube's rule and using tabulated data for amino acids and other small molecules or groups [16, 17]. This method seems to yield results of sufficient accuracy also with pigment-containing membrane proteins (H. Durchschlag, personal communication). Naturally, it is only applicable when the exact composition of the protein/pigment complex is known.

Edelstein and Schachman have suggested to perform equilibrium sedimentation at two different solvent densities, ρ_{01} and ρ_{02}, and to different

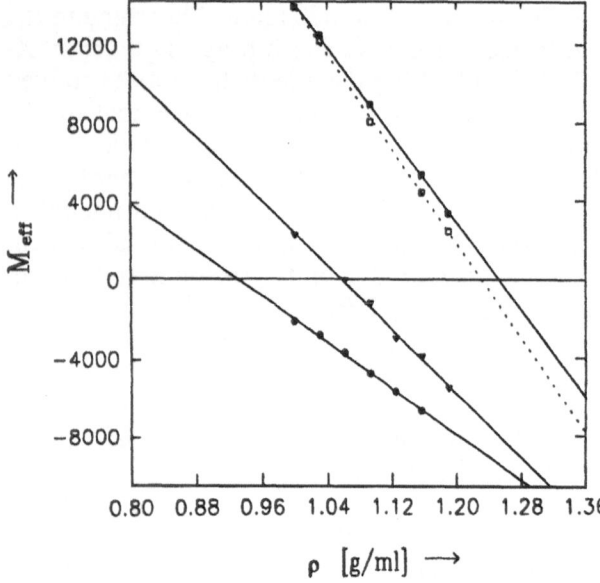

Fig. 1. Dependency of the effective molar mass of detergent micelles on solvent density, in $H_2O/D_2O/D_2{}^{18}O$ mixtures. (●): LDAO, (▼) $C_{12}E_9$, (■) $C_{12}M$. The samples contained approx. 1 mg/ml of the respective detergent and 1 μg/ml DPH. The data yield the following detergent densities (in g/ml): LDAO, 0.933; $C_{12}E_9$, 1.059; $C_{12}M$, 1.253. The "true" density of $C_{12}M$, as corrected for H-D exchange ((□) and dotted line), is 1.229 g/ml. The corresponding relative micelle masses, per mol, are 27800, 43000, and 75000, respectively. The density increments due to added salts (up to 0.1 M) were found to be additive and can easily be corrected for. The contribution of the dye to the micelle density is negligible[+]

both M and $\bar{v}$ of the particles studied by combining the results of both runs:

$$M(1 - \bar{v}\rho_{01}) = A, \quad M(1 - \bar{v}\rho_{02}) = B$$

(where A and B are the respective effective molar masses found) [6]. By applying this method, even protein/pigment complexes of unknown composition can be studied. To study intrinsic membrane proteins in detergent solution, however, the method has to be adapted. Two prerequisites have to be fulfilled: (1) The protein must be available in at least two detergents the densities of which differ as much as possible, and (2) the state of association of the protein in the two detergents must be identical. When this is the case, either sedimentation equilibrium runs have to be performed at solvent densities which match the buoyant densities of the respective detergents, or the $M(1 - \bar{v}\rho_0)$-data collected at differing solvent densities have to be extrapolated to those density values. In contrast to the original Edelstein-and-Schachman method, the solvent densities used in the calculation are thus not arbitrary but fixed by the properties of the respective detergent. A graph of the figures thus determined versus solvent density appears to be useful (see below).

Correcting for H-D exchange

As described by Edelstein and Schachman, the results obtained by their method are affected by replacement of exchangeable hydrogen atoms in the protein molecule by deuterium atoms. Due to this effect, the slope of the straight lines in a ln c-versus-r^2 plot resulting from a sedimentation equilibrium run relates, not to $M(1 - \bar{v}\rho_0)$, but to $M(1 + \Delta k_p - \bar{v}\rho_0)$, where $(1 + \Delta k_p)$ is the ratio of the protein's molar mass in the (partly) deuterated to that in the nondeuterated solvent. For a variety of water-soluble proteins, the maximum value of Δk_p, $\Delta k_{p,\,max}$, as obtained in 100% D_2O, was found to be 0.0155; when the D_2O content of the solvent is lower, Δk_p is reduced proportionately [6]. However, intrinsic membrane proteins embedded in detergent micelles most probably contain a much smaller percentage of exchangeable hydrogen ions than water-soluble ones. We therefore expect the average $\Delta k_{p,\,max}$-value for these proteins to reach only approx. half the value given above (i.e. 0.008). The influence of the deuterium exchange on the final results is therefore small. It can be corrected for (in a first but usually sufficient approximation) by subtracting, from the M_{eff}-values obtained, $M'\Delta k_p$, where M' is the molar mass calculated from the data without considering H-D exchange, and then proceeding with the corrected figures for M_{eff}. If necessary,

---

[+] The densities determined for LDAO and $C_{12}E_9$ differ from those used in a recent paper from our group on the light-harvesting complex II (B 800/850) from Rhodospirillum molischianum [9]. Use of the revised figures would reduce the molar masses derived in [9] by 8.0% (LDAO) and approx. 2.0% ($C_{12}E_9$), respectively. Correcting for H-D exchange (see below) would lead to a further reduction of M in $C_{12}E_9$ by 2.6% but would increase the resulting figure in LDAO by 3.0% ($\Delta k_{p,\,max} = 0.008$). The conclusion on the octameric structure of LH II drawn in [9] would not be affected.

another cycle of correction, using the improved values for M', could be performed.

Some detergents, in particular $C_{12}M$, also contain exchangeable hydrogen atoms ($\Delta k_{d, max}$ for $C_{12}M$ is 0.014). Two consequences arise from the exchange: (1) In the determination of detergent density, analogous corrections have to be applied as with proteins (using the uncorrected molar mass of the detergent micelle which can easily be obtained from the data of Fig. 1). (2) Besides the increase, due to H-D exchange, of the protein molar mass, also the corresponding increase in the molar mass M_d of the bound detergent has to be considered:

$$M_{eff} = M(1 + \Delta k_p - \bar{v}\rho_0)$$

$$+ M_d(1 + \Delta k_d - \bar{v}_d\rho_0)$$

M_d can be determined, again in a first but usually sufficient approximation, from the difference in

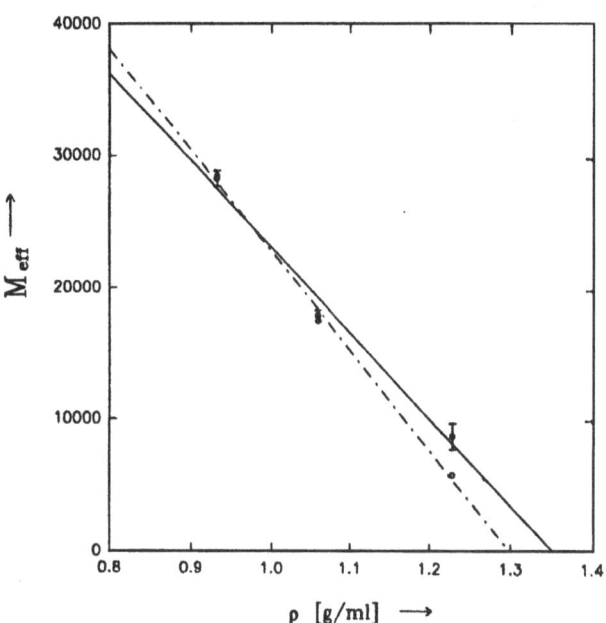

Fig. 3. Dependency of the effective molar mass of LH III, under conditions of density matching, on solvent density. The open symbols and the dotted line represent the data as corrected for H-D exchange

the uncorrected molar masses of the protein/pigment/detergent and the protein/pigment complex. The former mass can be derived from a plot of M_{eff} versus solvent density (together with the corresponding partial specific volume, cf. Fig. 2). The determination of the latter is described above (cf. Fig. 3)*.

Oligomeric structure of LH III from Rhodospirillum molischianum

According to electrophoresis under non-denaturing conditions, LH III from Rhodospirillum molischianum in detergent solution is an oligomer of a protein/pigment unit composed of two types of polypeptides of only partially known sequence, bacteriochlorophyll a and a carotenoid. The molar mass of this unit is expected to be 14800–15900 [8]. Recent electronmicroscopic studies on the oligomer, also in detergent solution,

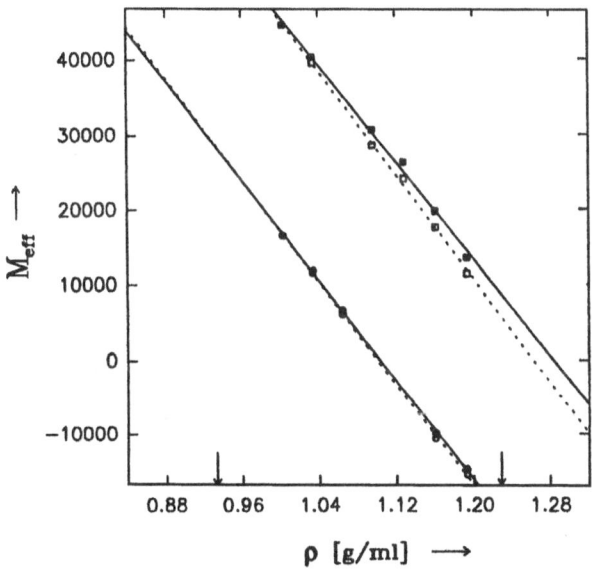

Fig. 2. LH III in solutions of LDAO (●) and $C_{12}M$ (■): Dependency of the effective molar mass on solvent density. The arrows indicate the buoyant density of the respective detergent according to Fig. 1. The open symbols and the dotted lines are obtained when the data are corrected for H-D exchange

---

*For the experiments described below, the error in $\bar{v}$ and M resulting from the use of uncorrected values for the detergent density and the molar mass of the protein-bound detergent is below 1%. This is due to a partial cancellation of the different contributions.

suggest it to be a hexamer of the basic unit; a pentameric or even tetrameric structure could, however, not be fully excluded [18]. We have, therefore, applied sedimentation equilibrium analysis to this problem, using the method discussed above.

LH III in $C_{12}E_9$ was studied under conditions of density matching ($\rho_0 = 1.059$ g/ml, see Fig. 1). The particles were found to be homogeneous both in ln c-versus-r^2 plots and in fits to the experimental data using a single exponential [9]. The $M(1 - \bar{v}\rho_0)$-value for the complex was found to be 18000 ± 400 ($n = 12$) (without correction for H-D exchange).

In solutions both of $C_{12}M$ and LDAO, extrapolation of the $M(1 - \bar{v}\rho_0)$-data obtained at different solvent densities to the density of the detergent had to be applied. Both the uncorrected data and those corrected for H-D exchange are shown in Fig. 2; for the measurements in $C_{12}M$, the corrections include those for the protein-bound detergent. A plot of the M_{eff}-values extrapolated to the respective detergent densities, together with the corresponding figure for LH III in solutions of $C_{12}E_9$, as a function of solvent density is shown in Fig. 3. The data follow a straight line, as to be expected when the oligomeric structure of the protein/pigment particles is the same in all three detergents used.

The information sought, M and $\bar{v}$ of the protein-pigment complex, can be directly derived from Fig. 3: (1) From the point of intersection of the straight line with the abscissa, $\bar{v}$ is obtained. (2) Using the result for $\bar{v}$, M can be derived from any $M(1 - \bar{v}\rho_0)$-figure on the straight line. Fig. 3 thus yields (for $\Delta k_p = 0.008$): $\bar{v} = (0.769 \pm 0.010)$ ml g^{-1} and $M = 99000 \pm 6000$. For $\Delta k_{p, max} = 0.0155$ [6], $\bar{v} = 0.776$ ml g^{-1} and $M = 102000$ would be obtained. As discussed above, we consider the lower $\Delta k_{p, max}$-value as the more probable. The corresponding figure for M is the (6.5 ± 0.5)-fold of the relative molar mass of the monomeric α/β/pigment unit.

Additional data on LH III available from Fig. 2 are the partial specific volume and molar mass of the complex of protein, pigment, and LDAO or $C_{12}M$, respectively. When these data are combined with those for the protein/pigment complex, it follows that 1 g of the latter binds (0.9 ± 0.1)g of LDAO or (1.2 ± 0.1)g of $C_{12}M$. The detergent/protein ratio thus is distinctly

higher than observed with most other membrane proteins [3, 4].

Discussion

In the present paper, we have described an adaptation of the method of Edelstein and Schachman [6] for the simultaneous determination of the partial specific volumes and molar masses to intrinsic membrane proteins in detergent solutions. With pigment-containing membrane proteins of unknown or poorly known composition, this variant seems to be the only reliable procedure available. Fig. 3 clearly shows the importance of a yet unmentioned aspect: The density of one of the detergents should be near to that of the complex. It is obvious from the figure that, with protein/pigment complexes, $C_{12}M$ is a good choice for such a detergent.

With respect to the application of the methodology described to the oligomeric state of the light-harvesting complex B800/820 (LH III) from Rhodospirillum molischianum, our ultracentrifugation data do not fully settle the problem. According to our results, LH III can be either a hexamer or a heptamer of the α/β/pigment unit. On the other hand, a heptameric state of LH III can be ruled out by the following arguments: (1) It is not compatible with the results of electronmicroscopic studies, which point at the hexamer as the largest possible oligomer [18]. (2) It is barely compatible with the tendency of the particles to form crystals with triangular surface planes (L. Germeroth and H. Michel, unpublished data). We thus conclude that LH III is a hexamer of the basic unit.

Acknowledgements

We are grateful to R. Lancaster for critically reading the manuscript, and to the Deutsche Forschungsgemeinschaft (SFB 169) and the Fonds der Chemischen Industrie for financial support.

References

1. Deisenhofer J, Epp O, Miki K, Huber R, Michel H (1985) Nature 318:618–624.
2. Drews G (1985) Microbiol Rev 49:59–70.
3. Helenius A, Simons K (1975) Biochim Biophys Acta 415:29–79.

4. Tanford C, Reynolds JA (1976) Biochim Biophys Acta 457:133–177

5. Schubert D, Schuck P (1991) Progr Colloid Polym Sci 86:12–22.

6. Edelstein SJ, Schachman HK (1967) J Biol Chem 242:306–311.

7. Germeroth L, Lottspeich F, Robert B, Michel H (1993) Biochemistry 32:5615–5621.

8. Germeroth L (1993) Ph.D. Thesis, Frankfurt University.

9. Kleinekofort W, Germeroth L, van den Broek JA, Schubert D, Michel H (1992) Biochim Biophys Acta 1140:102–104.

10. Schubert D, Boss K, Dorst HJ, Flossdorf J, Pappert G (1983) FEBS Lett 163:81–84.

11. Schubert D, Boss K (1985) Z Naturforsch 40c: 908–911.

12. Kupke DW (1973) In: Leach SJ (ed) Physical Principles and Techniques of Protein Chemistry, Part C. Academic Press, New York, pp 1–75.

13. Durchschlag H, Jaenicke R (1982) Biochem Biophys Res Commun 108:1074–1079.

14. Kratky O, Leopold H, Stabinger H (1973) Methods Enzymol 27:98–110.

15. Butler PJG, Kühlbrandt W (1988) Proc Natl Acad Sci USA 85:3797–3801.

16. Cohn EJ, Edsall JT (1943) In: Cohn EJ, Edsall JT (eds) Proteins, Amino Acids and Peptides as Ions and Dipolar Ions. Hafner Publ Comp, New York, pp 370–381.

17. Durchschlag H (1986) In: Hinz H-J (ed) Thermodynamic Data for Biochemistry and Biotechnology. Springer, Berlin, pp 45–128.

18. Boonstra AF, Germeroth L, Boekema EJ (1994) Biochim Biophys Acta, in press.

Received June 18, 1993
accepted August 11, 1993

Author's address:

Prof. Dr. D. Schubert
Institut für Biophysik der J.W. Goethe-Universität
Theodor-Stern-Kai 7
Haus 74
60590 Frankfurt/M., FRG

Progress in Colloid & Polymer Science   Progr Colloid Polym Sci 94:20–39 (1994)

Calculation of the partial volume of organic compounds and polymers

H. Durchschlag[1]) and P. Zipper[2])

[1]) Institute of Biophysics and Physical Biochemistry, University of Regensburg, FRG
[2]) Institute of Physical Chemistry, University of Graz, Austria

Abstract: A novel, universal and easy-to-use approach is presented which allows the *ab initio* calculation of partial volumes of organic compounds, valid for aqueous solutions at 25 °C. The method is based on Traube's additivity principle and concept of volume increments for atoms. His concept, however, was improved considerably by manifold adaptations, corrections, and completions. Major improvements were performed in context with the increments for nitrogen, and the decrements for ring formation. Moreover, a procedure was developed for linking tabulated volumes of inorganic ions to calculated volumes of organic residues. The validity of our approach was confirmed by a systematic comparison of calculated and experimental volumes of different classes of organic and biochemical compounds, including small molecules and polymers of nonionic and ionic nature. The vast majority of calculated volumes is within a range of $\pm 2\%$, if compared with the experimental values. A detailed table summarizing calculated and experimental partial volumes of diverse organic and biochemical molecules is presented and allows estimation of errors to be expected when performing calculations of substances of unknown volume. The table may also be used as a powerful database for many purposes. The prediction of partial volumes of compounds of quite different structure and composition is of great importance for many fields of research: correct interpretation of the results from ultracentrifugal and solution scattering studies, statements on various solute–solvent interactions, etc.

Key words: Partial molar volume – partial specific volume – *ab initio* calculation – organic compounds – biochemical model compounds – polymers – ionic compounds – aqueous solution

1. Introduction

For a variety of physico-chemical techniques volumetric properties of small molecules and/or macromolecules are of importance. For some techniques the absolute values of the partial specific volume are required: for example, in analytical ultracentrifugation for determination of molar masses of macromolecules or macromolecule-ligand complexes, and for correction of sedimentation coefficients to standard conditions; in small-angle scattering for determination of molar masses and related quantities (mass per unit length or unit area). Some other methods take for granted constancy of the partial specific volume for the evaluation of data, e.g., size-exclusion chromatography and gel electrophoresis in the presence of sodium dodecyl sulfate.

The investigation of complex macromolecular structures (e.g., enzymes liganded with substrates, coenzymes, products and/or analogs; proteins associated with lipids, pigments and/or detergents) necessitates knowledge of the partial specific volume of both macromolecules and low-molecular ligands.

Partial specific volumes can be determined experimentally (e.g., by density measurements). Frequently, however, the experimental determination of the partial specific volume of macromolecules and/or low-molecular ligands is not feasible (e.g., due to scanty amounts of material, insufficient purity, uncertainties in sample concentration, handling problems such as instability or adsorption). In such cases, calculations or reliable estimations of volume quantities are required.

Calculations of partial specific or molar volumes reported in the literature make use of an additivity principle, first suggested by Kopp [1, 2], and later on modified by Traube [3, 4], and many others (e.g., [5–7]). This study comments on various approaches of this kind and presents a new and universal concept and volume increments which may be used for the *ab initio* calculation of volumes in aqueous solution. The validity of our approach is confirmed by comparing a great number (about 500) of calculated and experimental volumes of simple and complex organic molecules and polymers of different nature. A forerunner of the present approach has been presented by us for the case of detergents and lipids [8].

2. Calculation of partial volumes

2.1. Definitions

The partial specific volume, $\bar{v}_i$, of the ith component of a solution is defined as the change in total volume, ∂V, per unit mass upon adding an infinitesimal amount, ∂g_i, of component i at constant temperature, T, and pressure, P, and masses in grams, g_j, of all other components j:

$$\bar{v}_i = (\partial V/\partial g_i)_{T,P,g_j} \quad (j \neq i) . \tag{1}$$

The partial molar volume, $\bar{V}$, is defined in an analogous way by substituting the number of grams, g, by the number of moles, n:

$$\bar{V}_i = (\partial V/\partial n_i)_{T,P,n_j} \quad (j \neq i) . \tag{2}$$

Usually, partial specific volumes, $\bar{v}$, are given in cm^3/g, and partial molar volumes, $\bar{V}$, in cm^3/mol. These two quantities are related by:

$$\bar{v}_i = \bar{V}_i/M_i , \tag{3}$$

where M_i is the molar mass of the ith component, in g/mol.

In general, experimental values of partial volumes are determined by densimetry, and today, preferably by measurement in digital densimeters (cf. [9]). Measurements require exact knowledge of solute concentration. Furthermore, a strong concentration dependence of observed volumes

has to be considered in the case of ionic solutes (cf. [10]). For that reason volumes have to be extrapolated to concentration zero. For single determinations of volumes errors of about $\pm 1\%$ are usual.

For further details see [9, 11].

2.2. Survey of calculation procedures

As early as 1839, Kopp [1] suggested the concept of molar volume as an additive function of atomic volumes, and already proposed volume increments for some atoms present in organic molecules. The additivity for the case of inorganic electrolytes was established by many authors at the end of the last century (e.g., [12]).

Obvious deficiencies of Kopp's approach for organic compounds were eliminated by Traube [3, 4] by introducing corrections for covolume, ring formation, and ionization. From a critical comparison of experimentally observed molar volumes of various compounds of different nature, Traube derived increments for a variety of atoms. He obtained slightly different values for molecules in aqueous solution and for homogeneous substances, respectively.

Though Traube's additivity principle aimed at small organic molecules, the approach may also be applied to macromolecules (the concept of which was not known at that time). The most prominent example of this kind is the widely used method of Cohn and Edsall [13] for calculating the partial specific volume of proteins from the volume increments for the individual amino acid residues. Slight modifications of the Cohn–Edsall method were suggested by Zamyatnin [14, 15] and Perkins [16]. The additivity principle was also applied to carbohydrates [16], nucleic acids [17], and polymers in the solid state [18]. Calculations of partial specific volumes of detergents and lipids, starting from the known experimental volume of a homologous compound and adding the volume increment of the methylene group, were reported by Tanford, Reynolds, and co-workers [19, 20].

The additivity principle was applied by many authors to several problems. In general, authors did not use the increments given by Traube, but tried to introduce other increments for atoms and groups, based on quite different considerations: use of van der Waals volumes [21–27], volumes derived from space-filling models [21], volumes derived from crystal data [16, 28], derivation of consensus volumes [16]; volume increments for groups were obtained from manifold comparisons of volumes of different homologous series (e.g., [5, 29–48]); least-squares procedures applied to an extensive set of nonionic organic compounds of different nature yielded increments for diverse functional groups and correction parameters for bi- and polyfunctional compounds [6].

A comparison of the different approaches, however, showed that no general improvement of calculations could be achieved, at least when considering the whole set of organic molecules. Conceptually, the van der Waals volumes were found to be too small, unless further corrections were applied. In the case of proteins, volumes predicted from crystal structures delivered overestimates, when compared to direct densimetric data [16].

In this context, one has to stress that many tabulated values for increments are contradictory; they are only valid for special compounds and conditions. Moreover, some tabulated values are erroneous and many increments necessary for calculations are missing (e.g., [20]). Since obviously the different data sets published are not compatible, one should avoid mixing of increments from different calculation schemes. Special care is also advisable when increments for ions, not available in the tables mentioned, have to be taken from other literature sources (e.g., [10, 49–51]).

The objective of this study was to elaborate a procedure for the *ab initio* calculation of the partial volumes of organic and biochemical compounds in aqueous solution. Such a procedure should meet the following demands: it should be simple, easy to use, fast, accurate, and universally applicable (both to low- and high-molecular weight molecules, both nonionic and ionic). The most appropriate approach in this context appeared to be the concept of Traube [3, 4]. The generally cited literature sources to Traube's increments are the famous articles by Cohn and Edsall [13, 52]. Unfortunately, in these papers only a few increments (for C, H, N, O, S, and a few groups present in peptides) have been presented. Since this was an insufficient base for our purpose, we had to turn to the original papers of Traube. Though these papers contained a wealth of in-

formation, it soon turned out that a number of appropriate adaptations, corrections, and completions of the initial concept were necessary to enable comprehensive calculations.

2.3. Outline of the used calculation procedure

Ab initio calculations of partial molar volumes of diverse organic and biochemical compounds were performed following directions similar to those given by Traube [3, 4]. The partial molar volume in dilute aqueous solution was calculated from the volume increments for atoms and/or atomic groups, assuming additivity and allowing corrections for covolume, ring formation, and ionization:

$$\bar{V}_c = \Sigma V_i + V_{CV} - \Sigma V_{RF} - \Sigma V_{ES} , \qquad (4)$$

where V_i is the volume increment for any atom or atomic group, V_{CV} is the correction due to the covolume, V_{RF} and V_{ES} take into account the decrease of volume caused by ring formation and ionization (electrostriction), respectively. In the case of polymers the contribution of V_{CV} has to be neglected if the volume of the monomeric unit is to be calculated.

In a few cases the volume increments derived by Traube could be used (e.g., the values for C and H, cf. Table 1). In other cases, it was necessary to modify and/or complement the original data. The volume increments were chosen in such a manner that optimum coincidence between experimental and calculated volumes of representative model compounds was obtained. For this purpose, we had to consider a broad spectrum of different classes of compounds (homologous series, diverse derivatives).

The atomic increments for O could be taken over from Traube with only slight modifications and completions. For N, however, it was necessary to derive a gamut of new increments, markedly different from the few original values given by Traube [3, 4]. We also established a universal concept of decrements for ring formation, applicable to both homo- and heterocycles of different ring size.

In context with the volume calculations of diverse compounds, some increments may be missing in Table 1. The atomic volumes of some elements, necessary for calculation of some biochemically significant compounds, can be found

Table 1. Partial molar volume increments for atoms, valid for aqueous solutions at 25 °C.

Increments for Atoms	V_i (cm^3/mol) [a]
C	9.9[b][c]
H	3.1[b]
O (2nd or further neighboring OH)	0.4[b]
(OH or O$^-$ in a carboxylic group)	0.4[b]
(OH or O$^-$ in a phosphate, sulfate, or sulfonate)	0.4
(2nd =O at a ring, provided a tautomerism is possible)	0.9
(1st OH)	2.3[b]
(O$^-$ in a phenolate)	2.3
(−O− in an ether, ester, or anhydride)	5.5[b]
(O in a ring)	5.5
(=O, except 2nd =O at a ring)	5.5[d]
(O in an amine oxide or nitro group)	5.5
N (primary N in diamides)	0.5
(primary N in monoamides)	2.0
(diamines, polyamines, amine oxides, unsubstituted urea)	2.0
(α- and ω-amino groups in aminoacids and peptides)	2.0[e]
(amines at hydrated rings)	2.0
(primary, secondary, and tertiary monoamines; aminoalcohols)	4.0
(secondary and tertiary N in mono- and diamides)	4.0
(N in substituted ureas)	4.0
(2nd ammonium N)	4.0
(NH in hydrated rings)	4.0
(N without H in hydrated rings)	7.0
(N or NH in aromatic rings)	7.0
(nitro group and nitriles)	7.0
(guanidinium)	8.0[f]
(1st ammonium N, betaines)	12.2
S	15.5[b][g]
P (phosphines)	17.0[b]
(phosphonium)	28.5[b]
F	5.0[b]
Cl	15.0
Br	19.7
I	32.1

Special Increments and Decrements	V_j (cm^3/mol)
Covolume (V_{CV})	12.4[b]
Ring formation (V_{RF})	
3-membered ring	2.1[h]
4-membered ring	4.1[h]
5-membered ring	6.1[h]
6-membered ring	8.1[b][h]
7-membered ring	10.1[h]
8-membered ring	12.1[h]
≧ 9-membered ring	14.1[h]
Ionization (V_{ES})	
Contribution for one positive charge ($V_{ES}+$)	6.8[i]
Contribution for one negative charge ($V_{ES}-$)	6.7[i]

[a] The values mentioned do not contain the contribution of ionization, if any.
[b] Value already given by Traube [4].
[c] The same value holds for single-, double-, or triple-bonded C.
[d] Traube [4] used this value for all double-bonded O atoms.
[e] The value does not apply to the ring-N in Pro.
[f] The value refers to each N in a guanidinium group, in Arg only to the two terminal N.
[g] Traube's value for S [4] is identical to the atomic volume of S [53].
[h] The value may also be applied to heterocycles containing N and/or O.
[i] The sum of $V_{ES}+$ and $V_{ES}-$ equals 13.5, the value given by Traube [4] for one pair of monovalent ions.

Table 2. Atomic volumes of some selected elements at 25 °C.

Atomic Volumes	V_A (cm³/mol)[a]
Se	16.5
Mg	14.0
Ca	26.0
Mn	7.4
Fe	7.1
Co	6.6
Ni	6.6
Cu	7.1[b]
Zn	9.2
Mo	9.4
Hg	14.8

[a]) Values were taken from the compilation in [53].
[b]) Value already given by Traube [4].

Table 3. Partial molar volume increments for inorganic ions, valid for aqueous solutions at 25 °C.

Increments for Inorganic Ions	V_{ion} (cm³/mol)[a]
H^+	-3.8
Li^+	-4.7
Na^+	-5.0[b]
K^+	5.2
Cs^+	17.5
NH_4^+	14.6
Mg^{2+}	-28.8
Ca^{2+}	-25.4
Mn^{2+}	-25.3
Fe^{2+}	-32.3
Cu^{2+}	-31.8
Zn^{2+}	-29.2
Fe^{3+}	-55.1
OH^-	-0.2
F^-	2.6
Cl^-	21.6
Br^-	28.5
I^-	40.0
HCO_3^-	27.2
NO_3^-	32.8
$H_2PO_4^-$	32.9
CO_3^{2-}	3.3
HPO_4^{2-}	15.3
SO_4^{2-}	21.6

[a]) The values include already the contribution for ionization.
[b]) This value plus the contribution for one positive charge equals 1.8, the value given by Traube [4] for Na^+.

in Table 2 and in some compilations of physico-chemical data (e.g., [53]).

For the case of inorganic ions, the values, V_{ion}, presented in Table 3 may be used. These values were adapted from the compilation in [51]. In order to be compatible with the values in Table 1, we had to modify slightly the values, $V_{X(aq)}$, given in [51]:

$$\text{for cations: } V_{ion} = V_{X(aq)} + 1.6\,Z, \qquad (5)$$

$$\text{for anions: } V_{ion} = V_{X(aq)} - 1.6\,Z, \qquad (6)$$

where Z is the number of electric charges. It should be noted that, due to the shifts described in Eqs. (5) and (6), the volume increment for Na^+ given by Traube [4] and the increment in Table 3 differ by V_{ES+}.

The partial molar volume of an inorganic electrolyte may be obtained directly as sum of the corresponding values for cation and anion of Table 3. A slightly more sophisticated procedure, however, is required for combining the values in Table 3 with volumes of organic ions calculated according to Eq. (4) from values in Table 1.

The partial molar volume of an organic electrolyte, $\bar{V}_{c,el}$, may be calculated from:

$$\bar{V}_{c,el} = \Sigma \bar{V}_c + \Sigma V_{ion}^*. \qquad (7)$$

$\bar{V}_c$ in Eq. (7) is the contribution of the organic moiety of the electrolyte which may be obtained from Eq. (4). It should be stressed that $V_{ion}^* = V_{ion}$ in the case of inorganic cations, but $V_{ion}^* = V_{ion} - Z \times V_{CV}$ for inorganic anions.

It is plausible that in context with the different volume increments for some atoms (e.g., different values for O and N, depending on the state of bonding and/or neighboring groups) and the decrements for ring formation and ionization, the structural formula of the compound under investigation must be known.

In Table 4 volume increments for a variety of atomic groups of organic and biochemical compounds were summarized. These data were derived by us in order to illustrate the usage of the volume increments introduced in Table 1. Both the atomic and group volume increments, presented in Tables 1 and 4, may be used for calculating $\bar{V}_c$. For convenience, both sets of volumes may be mixed (thereby, however, keeping in mind the vicinity rules for O and N).

It should be noted that the partial volumes to be calculated, $\bar{V}_c$ and $\bar{v}_c$, are only valid for aqueous solution at 25 °C. Application of volumes at quite different temperatures (e.g., 4 °C) necessitates the use of a temperature correction. For substances in (aqueous) solution, temperature coefficients,

Table 4. Partial molar volume increments for some selected atomic groups, calculated from the data given in Table 1.

Increment	V_i (cm³/mol)
$-\overset{\mid}{\underset{\mid}{C}}H$	13.0
$-\overset{\mid}{C}H_2$	16.1
$-CH_3$	19.2
$-C_2H_5$	35.3
$-C_6H_4-$ (phenylene)	71.8[a]
$-C_6H_5$ (phenyl)	74.9[a]
$-C_6H_{10}-$ (cyclohexylene)	90.4[a]
$-C_6H_{11}$ (cyclohexyl)	93.5[a]
$-OH$ (1st OH)	5.4
$-OH$ (2nd or further neighboring OH)	3.5
$-\overset{\mid}{\underset{\mid}{C}}-OH$ (1st OH)	15.3
$-\overset{\mid}{\underset{\mid}{C}}-OH$ (2nd or further neighboring OH)	13.4
$-\overset{\mid}{\underset{OH}{C}}-\overset{\mid}{\underset{OH}{C}}-$	28.7
$-\overset{\mid}{\underset{OH}{C}}-\overset{\mid}{C}-\overset{\mid}{\underset{OH}{C}}-$	40.5
$-\overset{\mid}{\underset{OH}{C}}-\overset{\mid}{\underset{OH}{C}}-\overset{\mid}{\underset{OH}{C}}-$	42.1
$-O-\overset{\mid}{\underset{OH}{C}}-$	18.9
$-O-\overset{\mid}{C}-\overset{\mid}{\underset{OH}{C}}-$	30.7
$-\overset{O}{\overset{\|}{C}}-$ (except 2nd $=O$ at a ring)	15.4
$-\overset{O}{\overset{\|}{C}}-H$	18.5
$-\overset{O}{\overset{\|}{C}}-OH$	18.9
$-\overset{O}{\overset{\|}{C}}-O^-$	15.8[b]
$-\overset{O}{\overset{\|}{C}}-O-$ (ester, anhydride)	20.9
$-NH_2$ (monoamine, aminoalcohol)	10.2

Table 4. (Continued)

Increment	V_i (cm³/mol)
$-NH_2$ (diamine, polyamine, amine at hydrated ring, α- and ω-amino groups in amino-acids, monoamide, unsubstituted urea)	8.2
$-NH_2$ (diamide)	6.7
$-\overset{+}{N}H_3$ (1st ammonium)	21.5[b]
$\overset{+}{N}H_3$ (2nd ammonium)	13.3[b]
$-\overset{CH_3}{\underset{CH_3}{N^+}}-CH_3$ (1st ammonium)	69.8[b]
$-\overset{CH_3}{\underset{CH_3}{N^+}}-CH_3$ (2nd ammonium)	61.6[b]
$-\overset{O}{\overset{\|}{C}}-\overset{\mid}{N}-$ (tertiary N in monoamide, diamide, substituted urea)	19.4
$-\overset{O}{\overset{\|}{C}}-\overset{\mid}{N}H$ (secondary N in monoamide, diamide, substituted urea)	22.5
$-\overset{O}{\overset{\|}{C}}-NH_2$ (primary N in monoamide, unsubstituted urea)	23.6
$-\overset{O}{\overset{\|}{C}}-NH_2$ (primary N in diamide)	22.1
$-\overset{H}{\underset{}{N}}-\overset{H}{\underset{\mid}{C}}-\overset{O}{\overset{\|}{C}}-$ (peptide)	33.5
$-\overset{H}{\underset{\mid}{N}}-\overset{O}{\overset{\|}{C}}-\overset{H}{\underset{\mid}{N}}-$ (secondary N)	29.6
$-\overset{NH_2}{\underset{}{C}}=\overset{+}{N}H_2$ (guanidinium)	38.3[b]
$-C\equiv N$	16.9
$-\overset{\mid}{\underset{\mid}{N}}-O$ (amine oxide)	7.5
$-\overset{O}{\overset{\|}{N}}-O$ (nitro)	18.0
$-SH$	18.6
$-S-S-$	31.0
$-\overset{O}{\overset{\|}{S}}-$	21.0
$-\overset{O}{\overset{\|}{\underset{\|}{S}}}-O^-$ (with O below)	26.9[b]

Table 4. (Continued)

Increment	V_i (cm^3/mol)
$-O-\overset{\overset{O}{\|\|}}{\underset{\underset{O}{\|\|}}{S}}-O^-$	32.4[b]
$-\overset{}{\underset{\|}{P}}=O$ (phosphine oxide)	34.0
$-O-\overset{\overset{O}{\|\|}}{\underset{\underset{O^-}{\|}}{P}}-O-$	45.4[b]
$-O-\overset{\overset{O}{\|\|}}{\underset{\underset{O^-}{\|}}{P}}-O^-$	40.3[b]

[a]) The decrement for ring formation has to be taken into account separately.

[b]) The decrement for ionization has to be taken into account separately.

$\Delta \bar{v}/\Delta T$, of $2 - 10 \times 10^{-4}$ cm^3 g^{-1} K^{-1} have been reported in the literature; for volume corrections, therefore, a value of 5×10^{-4} cm^3 g^{-1} K^{-1} may be used [9].

3. Results and discussion

Calculated and experimental values for partial molar and specific volumes of different classes of representative organic and biochemical compounds in aqueous solution are presented in Table 5. The compounds presented comprise molecules exhibiting different characteristics: small molecules and polymers, both nonionic and ionic, mono- and polyfunctional molecules, aliphatic and aromatic compounds, homo- and heterocycles of different ring size, etc. We tried hard to cover the most important classes of compounds and their derivatives. The choice of compounds, however, is limited by the availability of experimental data. For comparison, a few examples for compounds in organic solvents are added.

The sequence of steps to be performed in the calculation procedure may be followed from the calculation examples outlined in Fig. 1. There is

no doubt that the most straightforward manner of acting only utilizes Tables 1–3, provided the user is aware of the underlying philosophy. This manner of proceeding is called "procedure 1" in Fig. 1. For most cases, use of Table 1 is already sufficient. Additional use of Tables 4 and 5 (procedures 2 and 3 in Fig. 1) seems to be more convenient, but requires consideration of some peculiar corrections (cf. ΔN and ΔO in the legend to Fig. 1).

As may be seen from a comparison of the calculated and experimental volumes of Table 5, the accordance of the volumes is satisfactory. About 75% of the calculated values do not exceed an error of $\pm 2\%$, and about 90% are within a range of $\pm 3\%$, if compared with the experimental volumes.

In a few cases more serious deviations may be found. Very pronounced deviations may be registered with the smallest representatives of homologous series (e.g., methane, acetaldehyde, sodium formate, sodium salts of ethanedioic acid, oxirane, ethanediamine chlorides), in contrast to the higher members of these families. Obviously, in these cases the increments (which have been derived from the set of homologous compounds) do not correctly match the behavior of the smallest members. Some substituted heterocycles may also exhibit atypical features (e.g., 1,3-dioxolan-2-one, 2,4,6-trimethyl-1,3,5-trioxane, 1-methyl-2-pyrrolidinone, 3-hydroxypyridine, cytidine). In a few cases the instability of the organic compound in aqueous solution may be responsible for the observed volume discrepancy (e.g., 1,3,5-triazine). Larger deviations between calculated and experimental volumes may also be found with some lipids (e.g., cholesterol, digitonin) and ionic detergents (alkyl sulfates and sulfonates). This may be attributed to special solute–solute and solute–solvent interactions of these substances. Solute–solvent interactions may also be responsible for the poorer accordance of volume data of polymers in the case of organic solvents. Last but not least, when comparing calculated and experimental volumes, one should also be aware of the possibility of slightly erroneous experimental values in some cases.

The partial molar volumes of solutes contain contributions of both the intrinsic volume (size of atoms) and volumetric effects due to solute–solvent interactions (e.g., hydration and hydrophobic effects). Therefore, the increments presented in

Table 5. Comparison of calculated and experimental partial volumes of some selected small molecules and polymers. Unless otherwise stated, the values are valid for aqueous solutions at 25 °C.

Molecules[a]	M (g/mol)	Calculated Volumes		Experimental Volumes[b]		Δ
		$\bar{V}_c$ (cm³/mol)	$\bar{v}_c$ (cm³/g)	$\bar{V}_{exp}$ (cm³/mol)	$\bar{v}_{exp}$ (cm³/g)	$\Delta\bar{V}$ (%)
1) Hydrocarbons:						
Methane	16.0	34.7	2.163	37.3 [6]	2.325	−7.0
Ethane	30.1	50.8	1.689	51.2 [6]	1.703	−0.8
Propane	44.1	66.9	1.517	67.0 [6]	1.519	−0.1
Fluoromethane	34.0	36.6	1.076	35.9 [6]	1.055	+1.9
Chloromethane	50.5	46.6	0.923	46.2 [6]	0.911	+0.9
Dichloromethane	84.9	58.5	0.689	58.1 [6]	0.684	+0.7
Trichloromethane (chloroform)	119.4	70.4	0.590	72.3 [6]	0.606	−2.6
Bromomethane	94.9	51.3	0.540	53.0 [6]	0.558	−3.2
Iodomethane	141.9	63.7	0.449	63.7 [6]	0.449	0
Bromoethane	109.0	67.4	0.619	66.7 [6]	0.612	+1.0
1-Bromopropane	123.0	83.5	0.679	82.2 [6]	0.668	+1.6
3-Bromo-1-propene	121.0	77.3	0.639	77.6 [6]	0.641	−0.4
Benzene	78.1	82.3	1.054	83.1 [6]	1.064	−1.0
Methylbenzene	92.1	98.4	1.068	97.7 [6]	1.060	+0.7
Nitrobenzene	123.1	97.2	0.790	97.7 [6]	0.794	−0.5
2) Monohydric Alcohols:						
Methanol	32.0	37.0	1.155	38.2 [7]	1.191	−3.1
Ethanol	46.1	53.1	1.153	55.1 [7]	1.196	−3.6
2-Propen-1-ol	58.1	63.0	1.085	64.3 [6]	1.107	−2.0
1-Butanol	74.1	85.3	1.151	86.6 [7]	1.168	−1.5
1-Hexanol	102.2	117.5	1.150	118.7 [7]	1.161	−1.0
2-Methoxyethanol	76.1	74.7	0.982	75.2 [7]	0.988	−0.7
2-Ethoxyethanol	90.1	90.8	1.008	91.2 [7]	1.012	−0.4
2-Propoxyethanol	104.2	106.9	1.026	107.1 [7]	1.028	−0.2
2-Butoxyethanol	118.2	123.0	1.041	123.0 [7]	1.041	0
Cyclobutanol	72.1	75.0	1.040	75.6 [7]	1.048	−0.8
Cyclopentanol	86.1	89.1	1.034	89.0 [7]	1.033	+0.1
Cyclohexanol	100.2	103.2	1.030	103.5 [7]	1.033	−0.3
Cyclooctanol	128.2	131.4	1.025	129.7 [7]	1.012	+1.3
Cyclopropanemethanol	72.1	77.0	1.068	76.0 [7]	1.054	+1.3
1-Cyclopropanethanol	86.1	93.1	1.081	92.4 [25]	1.073	+0.8
3) Diols and Polyols:						
1,2-Ethanediol	62.1	53.5	0.862	54.6 [7]	0.880	−2.0
1,2-Propanediol	76.1	69.6	0.915	71.2 [7]	0.936	−2.3
1,3-Propanediol	76.1	71.5	0.940	71.9 [7]	0.949	−0.6
1,4-Butanediol	90.1	87.6	0.972	88.3 [7]	0.980	−0.8
2,3-Butanediol	90.1	85.7	0.951	86.6 [7]	0.961	−1.0
1,6-Hexanediol	118.2	119.8	1.014	120.4 [7]	1.019	−0.5
1,8-Octanediol	146.2	152.0	1.039	152.6 [7]	1.044	−0.4
Diethylene glycol	106.1	93.1	0.877	92.2 [7]	0.869	+1.0
Triethylene glycol	150.2	130.8	0.871	129.3 [7]	0.861	+1.2
Tetraethylene glycol	194.2	168.5	0.868	166.3 [7]	0.856	+1.3
Dipropylene glycol	134.2	125.3	0.934	124.6 [6]	0.929	+0.6
Tripropylene glycol	192.3	179.1	0.932	177.6 [6]	0.924	+0.8
Tetrapropylene glycol	250.3	232.9	0.930	231.4 [6]	0.924	+0.6
1,2,3-Propanetriol (glycerol)	92.1	70.0	0.760	70.6 [6]	0.767	−0.8
meso-Erythritol	122.1	86.5	0.708	87.1 [7]	0.713	−0.7
Pentaerythritol	136.2	108.3	0.795	101.8 [6]	0.748	+6.4
cis-1,2-Cyclohexanediol	116.2	103.6	0.892	101.3 [7]	0.872	+2.3
trans-1,2-Cyclohexanediol	116.2	103.6	0.892	103.0 [7]	0.887	+0.6
1,4-Cyclohexanediol	116.2	105.5	0.908	105.3 [7]	0.907	+0.2

Table 5. (Continued)

Molecules[a]	M (g/mol)	Calculated Volumes		Experimental Volumes[b]		Δ
		$\bar{V}_c$ (cm^3/mol)	$\bar{v}_c$ (cm^3/g)	$\bar{V}_{exp}$ (cm^3/mol)	$\bar{v}_{exp}$ (cm^3/g)	$\Delta\bar{V}$ (%)
cis-1,5-Cyclooctanediol	144.2	133.7	0.927	132.5 [7]	0.919	+0.9
1,3-Adamantanediol	168.2	141.3	0.840	138.2 [7]	0.821	+2.2
1,4-Adamantanediol	168.2	141.3	0.840	139.3 [7]	0.828	+1.4
4) Phenols:						
Phenol	94.1	84.6	0.899	86.2 [7]	0.916	−1.8
Phenol, Na	116.1	69.8	0.601	66.3 [7]	0.571	+5.3
4-Nitrophenol	139.1	99.5	0.715	98.2 [7]	0.706	+1.3
4-Nitrophenol, Na	161.1	84.7	0.526	85.1 [7]	0.528	−0.5
1,2-Benzenediol	110.1	85.0	0.772	87.1 [7]	0.791	−2.4
1,3-Benzenediol	110.1	86.9	0.789	88.9 [7]	0.807	−2.2
1,4-Benzenediol	110.1	86.9	0.789	88.7 [7]	0.806	−2.0
5) Aminoalcohols:						
2-Aminoethanol	61.1	60.2	0.986	59.3 [7]	0.970	+1.6
3-Amino-1-propanol	75.1	76.3	1.016	75.2 [7]	1.001	+1.5
4-Amino-1-butanol	89.1	92.4	1.037	91.2 [6]	1.023	+1.3
6-Amino-1-hexanol	117.2	124.6	1.063	123.4 [6]	1.053	+1.0
2-(Methylamino)ethanol	75.1	76.3	1.016	77.1 [7]	1.026	−1.0
2-(Dimethylamino)ethanol	89.1	92.4	1.037	94.2 [7]	1.056	−1.9
2-(Ethylamino)ethanol	89.1	92.4	1.037	92.9 [7]	1.042	−0.5
2-(Diethylamino)ethanol	117.2	124.6	1.063	123.0 [7]	1.050	+1.3
2,2'-Iminobisethanol (diethanolamine)	105.1	94.7	0.901	94.4 [6]	0.897	+0.4
2,2'-(Ethylimino)bisethanol	133.2	126.9	0.953	125.0 [6]	0.939	+1.5
2,2',2''-Nitrilotrisethanol (triethanolamine)	149.2	129.2	0.866	127.8 [6]	0.857	+ 1.1
6) Ethers:						
1,1'-Oxybisethane (ethyl ether)	74.1	88.5	1.194	90.4 [7]	1.220	−2.1
Dimethoxymethane	76.1	77.9	1.024	80.4 [7]	1.057	−3.1
Diethoxymethane	104.2	110.1	1.057	113.9 [7]	1.094	−3.3
Tetramethoxymethane	136.2	121.1	0.889	126.2 [7]	0.927	−4.0
1,2-Dimethoxyethane	90.1	94.0	1.043	95.9 [7]	1.064	−2.0
1,2-Diethoxyethane	118.2	126.2	1.068	127.3 [7]	1.077	−0.9
1,1'-Oxybis(2-methoxyethane)	134.2	131.7	0.982	132.7 [7]	0.989	−0.8
1,1'-Oxybis(2-ethoxyethane)	162.2	163.9	1.010	165.0 [7]	1.017	−0.7
2,5,8,11-Tetraoxadodecane	178.2	169.4	0.950	169.7 [7]	0.952	−0.2
2,5,8,11,14-Pentaoxapentadecane	222.3	207.1	0.932	206.8 [7]	0.930	+0.1
7) Aldehydes:						
Acetaldehyde	44.1	50.1	1.137	43.7 [7]	0.992	+14.6
Benzaldehyde	106.1	97.7	0.921	96.1 [7]	0.905	+1.7
3-Hydroxybenzaldehyde	122.1	100.0	0.819	97.9 [7]	0.801	+2.2
3-Hydroxybenzaldehyde, Na	144.1	85.2	0.591	83.3 [7]	0.578	+2.3
4-Hydroxybenzaldehyde	122.1	100.0	0.819	96.9 [7]	0.794	+3.2
4-Hydroxybenzaldehyde, Na	144.1	85.2	0.591	83.5 [7]	0.580	+2.0
8) Ketones:						
2-Propanone (acetone)	58.1	66.2	1.140	66.9 [7]	1.152	−1.0
2-Butanone	72.1	82.3	1.141	82.5 [7]	1.144	−0.2
2-Pentanone	86.1	98.4	1.142	98.0 [7]	1.138	+0.4
2,5-Hexanedione	114.2	113.8	0.997	111.6 [6]	0.978	+2.0
Cyclobutanone	70.1	72.0	1.027	70.9 [7]	1.012	+1.6
Cyclohexanone	98.2	100.2	1.021	99.7 [7]	1.016	+0.5
Cyclooctanone	126.2	128.4	1.017	128.0 [7]	1.014	+0.3
Cyclononanone	140.2	142.5	1.016	142.0 [7]	1.013	+0.4
1,4-Cyclohexanedione	112.1	94.9	0.846	92.8 [6]	0.828	+2.3
2-Adamantanone	150.2	136.0	0.905	137.2 [6]	0.913	−0.9

Table 5. (Continued)

Molecules[a]		Calculated Volumes		Experimental Volumes[b]		Δ
	M	$\bar{V}_c$	$\bar{v}_c$	$\bar{V}_{exp}$	$\bar{v}_{exp}$	$\Delta \bar{V}$
	(g/mol)	(cm^3/mol)	(cm^3/g)	(cm^3/mol)	(cm^3/g)	(%)
9) Carboxylic Acids:						
Formic acid	46.0	34.4	0.747	34.7 [7]	0.754	−0.9
Formic acid, Na	68.0	19.6	0.288	25.1 [7]	0.368	−21.8
Acetic acid	68.1	50.5	0.742	51.9 [7]	0.763	−2.7
Acetic acid, Na	82.0	35.7	0.435	39.2 [7]	0.478	−8.9
Propanoic acid	74.1	66.6	0.899	67.9 [7]	0.917	−1.9
Propanoic acid, Na	96.1	51.8	0.539	53.8 [7]	0.560	−2.1
Butanoic acid	88.1	82.7	0.939	84.6 [7]	0.960	−2.2
Butanoic acid, Na	110.1	67.9	0.617	69.2 [7]	0.628	−1.9
Hexanoic acid	116.2	114.9	0.989	116.6 [7]	1.003	−1.4
Hexanoic acid, Na	138.1	100.1	0.725	101.1 [7]	0.732	−1.0
2-Hydroxyacetic acid	76.1	52.8	0.694	51.8 [7]	0.681	+1.9
2-Hydroxyacetic acid, Na	98.0	38.0	0.388	38.7 [7]	0.395	−1.9
2-Hydroxypropanoic acid	90.1	68.9	0.765	69.4 [7]	0.770	−0.7
2-Hydroxypropanoic acid, Na	112.1	54.1	0.483	55.1 [7]	0.492	−1.8
2-Hydroxybutanoic acid	104.1	85.0	0.816	85.5 [7]	0.821	−0.5
2-Hydroxybutanoic acid, Na	126.1	70.2	0.557	70.5 [7]	0.559	−0.4
2-Hydroxyhexanoic acid	132.2	117.2	0.887	117.3 [7]	0.888	−0.1
2-Hydroxyhexanoic acid, Na	154.1	102.4	0.664	102.2 [7]	0.663	+0.2
Ethanedioic acid	90.0	50.2	0.558	49.1 [7]	0.546	+2.2
Ethanedioic acid, Na	112.0	35.4	0.316	41.2 [7]	0.368	−14.1
Ethanedioic acid, Na$_2$	134.0	20.6	0.154	28.1 [7]	0.210	−26.7
Propanedioic acid	104.1	66.3	0.637	67.2 [7]	0.646	−1.3
Propanedioic acid, Na	126.0	51.5	0.409	56.0 [7]	0.444	−8.0
Propanedioic acid, Na$_2$	148.0	36.7	0.248	36.2 [7]	0.245	+1.4
Butanedioic acid	118.1	82.4	0.698	82.9 [7]	0.702	−0.7
Butanedioic acid, Na	140.1	67.6	0.483	68.9 [7]	0.492	−1.9
Butanedioic acid, Na$_2$	162.1	52.8	0.326	54.1 [7]	0.334	−2.4
Hexanedioic acid	146.1	114.6	0.784	115.7 [7]	0.792	−1.0
Hexanedioic acid, Na	168.1	99.8	0.594	101.0 [7]	0.601	−1.2
Hexanedioic acid, Na$_2$	190.1	85.0	0.447	86.2 [7]	0.453	−1.4
Tartaric acid	150.1	85.1	0.567	84.0 [7]	0.560	+1.3
Tartaric acid, Na	172.1	70.3	0.409	70.8 [7]	0.411	−0.7
Tartaric acid, Na$_2$	194.1	55.5	0.286	56.3 [7]	0.290	−1.4
10) Esters:						
Formic acid ethylester	74.1	71.7	0.968	72.8 [7]	0.983	−1.5
Acetic acid ethylester	88.1	87.8	0.996	88.9 [7]	1.009	−1.2
Ethanedioic acid dimethylester	118.1	92.6	0.784	91.4 [7]	0.774	+1.3
11) Heterocycles Containing 1 O:						
Oxirane [10 °C]	44.1	48.0	1.090	45.5 [6]	1.033	+5.5
Oxetane	58.1	62.1	1.069	61.4 [7]	1.057	+1.1
Tetrahydrofuran	72.1	76.2	1.057	76.9 [7]	1.066	−0.9
Tetrahydropyran	86.1	90.3	1.048	91.7 [7]	1.065	−1.5
Oxepane	100.2	104.4	1.042	105.5 [7]	1.053	−1.0
Tetrahydro-2-furanmethanol	102.1	94.6	0.926	93.8 [7]	0.918	+0.9
2,5-Dimethoxytetrahydrofuran	132.2	119.4	0.903	122.3 [7]	0.925	−2.4
2,5-Dimethoxy-2-(dimethoxymethyl)-tetrahydrofuran	206.2	178.7	0.866	177.1 [7]	0.859	+0.9
Dihydro-2(3H)-furanone (γ-butyrolactone)	86.1	75.5	0.877	73.3 [6]	0.851	+3.0
Dihydro-5-methyl-2(3H)-furanone (γ-valerolactone)	100.1	91.6	0.915	92.0 [6]	0.919	−0.4
Tetrahydro-2H-pyran-2-methanol	116.2	108.7	0.936	108.1 [7]	0.931	+0.6

Table 5. (Continued)

Molecules[a]		Calculated Volumes		Experimental Volumes[b]		Δ
	M (g/mol)	$\bar{V}_c$ (cm^3/mol)	$\bar{v}_c$ (cm^3/g)	$\bar{V}_{exp}$ (cm^3/mol)	$\bar{v}_{exp}$ (cm^3/g)	$\Delta\bar{V}$ (%)
12) Heterocycles Containing 2 or More O:						
1,3-Dioxolane	74.1	65.6	0.886	65.4 [7]	0.883	+0.3
1,3-Dioxane	88.1	79.7	0.905	80.8 [7]	0.917	−1.4
1,4-Dioxane	88.1	79.7	0.905	80.9 [7]	0.919	−1.5
1,3-Dioxepane	102.1	93.8	0.918	96.0 [7]	0.940	−2.3
1,3,5-Trioxane	90.1	69.1	0.767	69.5 [7]	0.772	−0.6
1,3-Dioxolan-2-one	88.1	64.9	0.737	62.3 [6]	0.707	+4.2
2,4,6-Trimethyl-1,3,5-trioxane	132.2	117.4	0.888	123.7 [7]	0.936	−5.1
1,4,7,10-Tetraoxacyclododecane (12-Crown-4)	176.2	149.1	0.846	149.9 [7]	0.851	−0.5
1,4,7,10,13-Pentaoxacyclopentadecane (15-Crown-5)	220.3	186.8	0.848	186.1 [7]	0.845	+0.4
1,4,7,10,13,16-Hexaoxacyclooctadecane (18-Crown-6)	264.3	224.5	0.849	223.4 [7]	0.845	+0.5
13) Monoamines:						
Methanamine	31.1	41.8	1.346	41.9 [7]	1.348	−0.2
Methanamine, Br	112.0	62.4	0.557	60.8 [54]	0.543	+2.6
Ethanamine	45.1	57.9	1.284	58.6 [7]	1.300	−1.2
Ethanamine, Br	126.0	78.5	0.623	77.7 [54]	0.616	+1.1
1-Butanamine	73.1	90.1	1.232	89.8 [7]	1.228	+0.3
1-Butanamine, Br	154.1	110.7	0.719	110.2 [54]	0.715	+0.5
1-Butanamine, NO$_3$	136.2	115.0	0.845	114.9 [47]	0.844	+0.1
1-Hexanamine	101.2	122.3	1.209	121.6 [7]	1.202	+0.6
1-Hexanamine, Br	182.1	142.9	0.785	142.0 [54]	0.780	+0.6
N-Methylmethanamine	45.1	57.9	1.284	59.2 [7]	1.313	−2.2
N-Methylmethanamine, Cl	81.6	71.6	0.878	72.5 [7]	0.889	−1.2
N-Ethylethanamine	73.1	90.1	1.232	91.7 [7]	1.254	−1.7
N-Ethylethanamine, Cl	109.6	103.8	0.947	106.7 [7]	0.974	−2.7
N-Butyl-1-butanamine	129.3	154.5	1.195	155.4 [7]	1.202	−0.6
N-Butyl-1-butanamine, Cl	165.7	168.2	1.015	170.7 [7]	1.030	−1.5
N,N-Dimethylmethanamine	59.1	74.0	1.252	79.0 [7]	1.336	−6.3
N,N-Dimethylmethanamine, Cl	95.6	87.7	0.918	90.6 [7]	0.948	−3.2
N,N-Diethylethanamine	101.2	122.3	1.209	120.9 [7]	1.195	+1.2
N,N-Diethylethanamine, Cl	137.7	136.0	0.988	138.6 [7]	1.007	−1.9
Tetramethylammonium bromide	154.1	110.7	0.719	114.3 [54]	0.742	−3.1
Tetraethylammonium bromide	210.2	175.1	0.833	173.6 [54]	0.826	+0.9
Tetrabutylammonium bromide	322.4	303.9	0.943	300.4 [54]	0.932	+1.2
Cyclopentanamine	85.2	91.9	1.079	91.0 [7]	1.069	+1.0
Cyclohexanamine	99.2	106.0	1.069	105.4 [7]	1.063	+0.6
Cyclooctanamine	127.2	134.2	1.055	133.4 [7]	1.048	+0.6
1-Adamantanamine	151.3	141.8	0.937	138.9 [7]	0.918	+2.1
Benzenamine (aniline)	93.1	89.4	0.960	89.3 [7]	0.959	+0.1
14) Diamines and Polyamines:						
1,2-Ethanediamine	60.1	61.0	1.015	62.7 [7]	1.043	−2.7
1,2-Ethanediamine, Cl	96.6	76.7	0.794	73.9 [7]	0.765	+3.8
1,2-Ethanediamine, Cl$_2$	133.0	84.2	0.633	80.0 [7]	0.601	+5.2
1,3-Propanediamine	74.1	77.1	1.040	78.4 [7]	1.058	−1.7
1,3-Propanediamine, Cl	110.6	92.8	0.839	90.7 [7]	0.820	+2.3
1,3-Propanediamine, Cl$_2$	147.1	100.3	0.682	98.5 [7]	0.670	+1.8
1,4-Butanediamine	88.2	93.2	1.057	93.6 [7]	1.062	−0.4
1,4-Butanediamine, Cl	124.6	108.9	0.874	107.5 [7]	0.863	+1.3
1,4-Butanediamine, Cl$_2$	161.1	116.4	0.723	116.3 [7]	0.722	+0.1
1,6-Hexanediamine	116.2	125.4	1.079	124.8 [7]	1.074	+0.5
1,6-Hexanediamine, Cl	152.7	141.1	0.924	139.8 [7]	0.916	+1.0

Table 5. (Continued)

Molecules[a]		Calculated Volumes		Experimental Volumes[b]		Δ
	M	$\bar{V}_c$	$\bar{v}_c$	$\bar{V}_{exp}$	$\bar{v}_{exp}$	$\Delta\bar{V}$
	(g/mol)	(cm^3/mol)	(cm^3/g)	(cm^3/mol)	(cm^3/g)	(%)
1,6-Hexanediamine, Cl$_2$	189.1	148.6	0.786	151.3 [7]	0.800	−1.8
1,8-Octanediamine	144.3	157.6	1.092	157.3 [7]	1.090	+0.2
1,8-Octanediamine, Cl	180.7	173.3	0.959	171.6 [7]	0.950	+1.0
1,8-Octanediamine, Cl$_2$	217.2	180.8	0.832	183.8 [7]	0.846	−1.6
1,10-Decanediamine	172.3	189.8	1.102	188.3 [7]	1.093	+0.8
1,10-Decanediamine, Cl	208.8	205.5	0.984	203.3 [7]	0.974	+1.1
1,10-Decanediamine, Cl$_2$	245.2	213.0	0.869	215.5 [7]	0.879	−1.2
trans-1,4-Cyclohexanediamine	112.2	104.9	0.935	108.9 [7]	0.971	−3.7
N,N′-Octamethylenebis(tributylammonium) dibromide	642.7	581.0	0.904	579.4 [54]	0.901	+0.3
Diethylenetriamine	103.2	98.3	0.953	101.2 [7]	0.981	−2.9
Triethylenetetramine	146.2	135.6	0.927	137.6 [7]	0.941	−1.5
Tetraethylenepentamine	189.3	172.9	0.913	175.9 [7]	0.929	−1.7
15) Monoamides:						
Formamide	45.0	39.1	0.868	38.5 [7]	0.855	+1.6
Acetamide	59.1	55.2	0.934	55.7 [7]	0.943	−0.9
Propanamide	73.1	71.3	0.975	71.5 [7]	0.978	−0.3
Butanamide	87.1	87.4	1.003	87.1 [7]	1.000	+0.3
Hexanamide	115.2	119.6	1.038	119.2 [7]	1.035	+0.3
N-Methylformamide	59.1	57.2	0.968	56.8 [7]	0.962	+0.7
N-Ethylformamide	73.1	73.3	1.003	74.0 [7]	1.012	−0.9
N-Propylformamide	87.1	89.4	1.026	87.9 [7]	1.009	+1.7
N-Methylacetamide	73.1	73.3	1.003	74.0 [7]	1.013	−1.0
N-Ethylacetamide	87.1	89.4	1.026	90.7 [7]	1.041	−1.4
N-Propylacetamide	101.2	105.5	1.043	105.1 [7]	1.039	+0.4
N-Methylpropanamide	87.1	89.4	1.026	89.8 [7]	1.030	−0.4
N-Ethylpropanamide	101.2	105.5	1.043	105.4 [7]	1.042	+0.1
N-Propylpropanamide	115.2	121.6	1.056	121.5 [7]	1.055	+0.1
N,N-Dimethylformamide	73.1	73.3	1.003	74.5 [7]	1.019	−1.6
N,N-Diethylformamide	101.2	105.5	1.043	106.7 [7]	1.055	−1.1
N,N-Dimethylacetamide	87.1	89.4	1.026	89.7 [7]	1.029	−0.3
N,N-Diethylacetamide	115.2	121.6	1.056	121.5 [7]	1.055	+0.1
N,N-Dipropylacetamide	143.2	153.8	1.074	154.2 [7]	1.077	−0.3
N,N-Bis(1-methylethyl)acetamide	143.2	153.8	1.074	152.0 [7]	1.061	+1.2
N,N-Dimethylpropanamide	101.2	105.5	1.043	105.4 [7]	1.042	+0.1
N,N-Diethylpropanamide	129.2	137.7	1.066	137.7 [7]	1.066	0
2-Hydroxyacetamide	75.1	57.5	0.766	56.2 [7]	0.749	+2.3
2-Hydroxypropanamide	89.1	73.6	0.826	73.3 [7]	0.823	+0.4
4-Hydroxybutanamide	103.1	89.7	0.870	88.9 [7]	0.862	+0.9
6-Hydroxyhexanamide	131.2	121.9	0.929	121.1 [7]	0.923	+0.7
16) Diamides:						
Ethanediamide	88.1	56.6	0.643	55.0 [7]	0.625	+2.9
Propanediamide	102.1	72.7	0.712	72.6 [7]	0.711	+0.1
Butanediamide	116.1	88.8	0.765	88.8 [7]	0.765	0
N,N,N′,N′-Tetramethyloctanediamide	228.3	224.6	0.984	224.0 [6]	0.981	+0.3
17) Urea and Guanidinium Salts:						
Urea	60.1	44.2	0.703	44.2 [7]	0.703	0
Methylurea	74.1	64.3	0.868	62.1 [7]	0.838	+3.5
Ethylurea	88.1	80.4	0.912	80.2 [7]	0.910	+0.2
Propylurea	102.1	96.5	0.945	94.9 [7]	0.929	+1.7
Butylurea	116.2	112.6	0.969	116.1 [7]	0.999	−3.0
(1-Methylethyl)urea	102.1	96.5	0.945	99.0 [7]	0.969	−2.4

Table 5. (Continued)

Molecules[a])	M (g/mol)	Calculated Volumes		Experimental Volumes[b])		Δ
		$\bar{V}_c$ (cm^3/mol)	$\bar{v}_c$ (cm^3/g)	$\bar{V}_{exp}$ (cm^3/mol)	$\bar{v}_{exp}$ (cm^3/g)	$\Delta\bar{V}$ (%)
N,N-Dimethylurea	88.1	80.4	0.912	78.9 [7]	0.895	+1.9
N,N'-Dimethylurea	88.1	80.4	0.912	80.0 [7]	0.908	+0.5
N,N-Diethylurea	116.2	112.6	0.969	114.9 [7]	0.989	−2.0
N,N'-Diethylurea	116.2	112.6	0.969	116.1 [7]	0.999	−3.0
Tetramethylurea	116.2	112.6	0.969	115.6 [7]	0.995	−2.6
Hydantoic acid	118.1	78.1	0.661	77.6 [13]	0.657	+0.6
Guanidinium chloride [20 °C]	95.5	67.3	0.704	66.5	0.696 [9]	+1.2
Propylguanidinium nitrate	164.2	126.8	0.772	127.1 [47]	0.774	−0.2
18) Nitriles:						
Ethanenitrile	41.1	48.5	1.181	47.4 [7]	1.155	+2.3
Propanenitrile	55.1	64.6	1.173	64.3 [7]	1.167	+0.5
Butanenitrile	69.1	80.7	1.168	80.6 [7]	1.166	+0.1
Hexanenitrile	97.2	112.9	1.162	111.9 [7]	1.152	+0.9
Octanenitrile	125.2	145.1	1.159	146.2 [7]	1.168	−0.8
4-Hydroxybenzonitrile	119.1	98.4	0.826	98.3 [7]	0.825	+0.1
4-Hydroxybenzonitrile, Na	141.1	83.6	0.592	84.1 [7]	0.596	−0.4
19) Heterocycles Containing 1 N:						
Aziridine	43.1	49.6	1.152	48.9 [7]	1.135	+1.4
Azetidine	57.1	63.7	1.116	63.7 [7]	1.116	0
Pyrrolidine	71.1	77.8	1.094	77.8 [7]	1.094	0
Piperidine	85.2	91.9	1.079	92.3 [7]	1.084	−0.4
Hexahydro-1H-azepine	99.2	106.0	1.069	105.6 [6]	1.064	+0.4
Octahydroazocine	113.2	120.1	1.061	120.1 [6]	1.061	0
Pyridine	79.1	76.3	0.965	77.4 [7]	0.979	−1.4
Pyridine, Cl	115.6	87.0	0.753	90.7 [7]	0.785	−4.1
1-Methyl-2-pyrrolidinone	99.1	96.2	0.970	90.4 [7]	0.912	+6.4
2,5-Pyrrolidinedione (succinimide)	99.1	71.8	0.725	72.3 [6]	0.730	−0.7
1-Methylpiperidine	99.2	111.0	1.119	110.5 [7]	1.114	+0.5
1-Methylpiperidine, Cl	135.6	121.7	0.897	125.5 [7]	0.925	−3.0
2-Piperidinone (δ-valerolactam)	99.1	91.2	0.920	90.3 [6]	0.911	+1.0
Hexahydro-1H-azepin-2-one (ε-caprolactam)	113.2	105.3	0.931	105.0 [6]	0.928	+0.3
4-Aminopyridine	94.1	83.4	0.886	82.7 [7]	0.879	+0.8
3-Hydroxypyridine	95.1	78.6	0.754	75.4 [7]	0.793	+4.2
4,4'-Bipyridine	156.2	134.0	0.858	133.2 [7]	0.853	+0.6
20) Heterocycles Containing 2 or More N:						
Pyrazole	68.1	62.4	0.917	62.3 [7]	0.915	+0.2
Imidazole	68.1	62.4	0.917	60.0 [7]	0.881	+4.1
Piperazine	86.1	82.9	0.962	83.4 [7]	0.968	−0.6
Pyrazine	80.1	70.3	0.878	70.8 [7]	0.884	−0.7
Pyrimidine	80.1	70.3	0.878	70.3 [7]	0.878	0
Pyridazine	80.1	70.3	0.878	70.4 [7]	0.879	−0.1
1,3,5-Triazine	81.1	64.3	0.793	54.8 [7]	0.676	+17.3
2,4-Imidazolidinedione (hydantoin)	100.1	62.8	0.627	65.0 [6]	0.649	−3.4
1-Methylpiperazine	100.2	102.0	1.018	102.3 [7]	1.021	−0.3
1-Methylpiperazine, Cl$_2$	173.1	118.2	0.683	117.5 [7]	0.679	+0.6
2,5-Piperazinedione	114.1	76.9	0.674	76.7 [6]	0.672	+0.3
1,4-Dimethyl-2,5-piperazinedione	142.2	115.1	0.810	113.4 [6]	0.798	+1.5
3,6-Dimethyl-2,5-piperazinedione	142.2	109.1	0.767	112.4 [6]	0.791	−2.9
1,4-Diazabicyclo[2,2,2]octane	112.2	106.8	0.952	105.9 [6]	0.944	+0.8
1,3,5,7-Tetraazatricyclo[3,3,1,1^{3,7}]decane (hexamethylenetetramine)	140.2	112.7	0.804	111.4 [6]	0.795	+1.2

Table 5. (Continued)

Molecules[a]	M (g/mol)	Calculated Volumes		Experimental Volumes[b]		Δ
		$\bar{V}_c$ (cm³/mol)	$\bar{v}_c$ (cm³/g)	$\bar{V}_{exp}$ (cm³/mol)	$\bar{v}_{exp}$ (cm³/g)	$\Delta\bar{V}$ (%)
21) Condensed Heterocycles Containing N:						
1H-Benzimidazole	118.1	100.1	0.847	98.5 [7]	0.834	+1.6
1H-Benzotriazole	119.1	94.1	0.790	94.5 [7]	0.793	−0.4
Quinoline	129.2	114.0	0.883	115.5 [7]	0.894	−1.3
Quinazoline	130.2	108.0	0.830	108.5 [7]	0.833	−0.5
Quinoxaline	130.2	108.0	0.830	109.2 [7]	0.839	−1.1
1,10-Phenanthroline	180.2	145.7	0.809	142.4 [7]	0.790	+2.3
Purine	120.1	88.1	0.733	85.4 [7]	0.711	+3.2
Caffeine	194.2	147.4	0.759	145.2 [7]	0.748	+1.5
22) Heterocycles Containing N and O:						
Morpholine	87.1	81.3	0.933	82.6 [7]	0.948	−1.5
4-Methylmorpholine	101.2	100.4	0.993	101.3 [7]	1.001	−0.9
4-Methylmorpholine, Cl	137.6	111.1	0.807	111.7 [7]	0.812	−0.5
23) Compounds Containing S or P:						
Benzenethiol (thiophenol)	110.2	97.8	0.888	94.1 [6]	0.854	+3.9
1,1′-Thiobisethane (thioether)	122.3	98.5	0.806	99.5 [6]	0.814	−1.0
Sulfinylbismethane (dimethylsulfoxide)	78.1	71.8	0.919	68.9 [6]	0.882	+4.2
Ethanethioamide (thioacetamide)	75.1	65.2	0.868	66.4 [6]	0.884	−1.8
Thiourea	76.1	54.2	0.712	54.8 [6]	0.720	−1.1
Hydrazinecarbothioamide (thiosemicarbazide)	91.1	65.3	0.716	64.0 [7]	0.702	+2.0
2-Thioxo-imidazolidinone (thiohydantoin)	116.1	77.4	0.666	76.5 [7]	0.659	+1.2
Tetraphenylphosphonium chloride	374.9	310.5	0.828	310.0 [55]	0.827	+0.2
24) Amino Acids, Betaines, and Peptides: Amino Acids:						
Glycine	75.1	42.1	0.561	43.3 [7]	0.577	−2.8
L-Alanine	89.1	58.2	0.653	60.5 [7]	0.679	−3.8
L-Valine	117.2	90.4	0.772	90.8 [7]	0.775	−0.4
L-Leucine	131.2	106.5	0.812	107.8 [7]	0.822	−1.2
L-Isoleucine	131.2	106.5	0.812	105.8 [7]	0.807	+0.7
L-Serine	105.1	60.5	0.576	60.6 [7]	0.577	−0.2
L-Threonine	119.1	76.6	0.643	76.9 [7]	0.646	−0.4
L-Cysteine	121.2	73.7	0.608	73.5 [7]	0.606	+0.3
L-Methionine	149.2	105.9	0.710	105.6 [7]	0.708	+0.3
L-Proline	115.1	80.1	0.696	82.8 [7]	0.719	−3.3
L-Phenylalanine	165.2	121.9	0.738	121.5 [7]	0.736	+0.3
Tyrosine	181.2	124.2	0.685	123.6 [40]	0.682	+0.5
DL-Tryptophan	204.2	145.7	0.713	143.9 [7]	0.704	+1.3
L-Aspartic acid	133.1	74.0	0.556	74.8 [7]	0.562	−1.1
L-Asparagine	132.1	78.7	0.596	78.0 [7]	0.590	+0.9
L-Glutamic acid	147.1	90.1	0.612	89.9 [7]	0.611	+0.3
Glutamine	146.2	94.8	0.649	93.9 [40]	0.642	+1.0
Lysine [20 °C]	146.2	111.6	0.763	109.0 [40]	0.746	+2.4
L-Arginine	174.2	127.6	0.732	127.4 [7]	0.731	+0.2
L-Histidine	155.2	102.0	0.657	98.3 [7]	0.634	+3.8
2-Aminobutanoic acid	103.1	74.3	0.721	75.6 [7]	0.733	−1.7
3-Aminobutanoic acid	103.1	76.3	0.740	76.3 [7]	0.740	0
4-Aminobutanoic acid	103.1	74.3	0.721	73.2 [7]	0.710	+1.5
2-Aminohexanoic acid (norleucine)	131.2	106.5	0.812	108.4 [13]	0.826	−1.8
6-Aminohexanoic acid	131.2	106.5	0.812	104.2 [7]	0.794	+2.2
8-Aminooctanoic acid	159.2	138.7	0.871	136.1 [7]	0.855	+1.9
Betaines:						
Betaine	117.2	100.6	0.859	97.7 [13]	0.834	+3.0

Table 5. (Continued)

Molecules[a]	M (g/mol)	Calculated Volumes		Experimental Volumes[b]		Δ
		$\bar{V}_c$ (cm^3/mol)	$\bar{v}_c$ (cm^3/g)	$\bar{V}_{exp}$ (cm^3/mol)	$\bar{v}_{exp}$ (cm^3/g)	$\Delta\bar{V}$ (%)
N-Dimethylphenylglycine	179.2	148.2	0.827	146.8 [13]	0.819	+1.0
Peptides:						
Diglycine	132.1	78.7	0.596	76.3 [7]	0.578	+3.1
Triglycine	189.2	115.3	0.610	112.2 [7]	0.593	+2.8
Tetraglycine	246.2	151.9	0.617	149.7 [7]	0.608	+1.5
Pentaglycine	303.3	188.5	0.622	187.1 [7]	0.617	+0.7
Dialanine	160.2	110.9	0.692	110.6 [7]	0.691	+0.3
Trialanine	231.2	163.6	0.708	163.8 [7]	0.708	−0.1
Tetraalanine	302.3	216.3	0.716	220.1 [7]	0.728	−1.7
Diserine	192.2	115.5	0.601	111.8 [7]	0.582	+3.3
Triserine	279.2	170.5	0.611	166.0 [7]	0.594	+2.7
Glycylalanine	146.1	94.8	0.649	92.7 [7]	0.634	+2.3
Alanylglycine	146.1	94.8	0.649	95.0 [7]	0.650	−0.2
Glycylalanylglycine	203.2	131.4	0.647	132.4 [47]	0.652	−0.8
Glycylserylglycine	219.2	133.7	0.610	133.1 [47]	0.607	+0.5
25) Nucleobases, Nucleosides, and Nucleotides:						
Hydrouracil	114.1	76.9	0.674	76.3 [6]	0.669	+0.8
Adenosine [temperature not specified]	267.3	173.9	0.651	170.5	0.638 [9]	+1.3
Cytidine	243.2	159.6	0.656	153.5 [7]	0.631	+4.0
Uridine	244.2	152.4	0.624	151.5 [7]	0.620	+0.6
Thymidine	242.2	168.5	0.696	167.6 [7]	0.692	+0.6
FMN, Na	478.4	281.5	0.588	274.6	0.574 [9]	+2.5
NAD$^+$, free acid [20 °C]	663.4	409.4	0.617	411.3	0.62 [9]	−0.5
NADH, Na$_2$ [20 °C]	709.4	394.3	0.556	404.4	0.57 [9]	−2.5
26) Carbohydrates:						
Ribose	150.1	95.8	0.638	95.2 [7]	0.634	+0.6
2-Deoxy-D-ribose	134.1	97.3	0.725	93.8 [7]	0.699	+3.7
D-Glucose	180.2	110.3	0.612	111.1 [7]	0.617	−0.7
2-Deoxy-D-glucose	164.2	111.8	0.681	110.4 [7]	0.673	+1.3
D-Fructose	180.2	110.3	0.612	110.4 [7]	0.613	−0.1
Sucrose (saccharose)	342.3	214.4	0.626	211.3 [7]	0.617	+1.5
D-Mannitol	182.2	119.5	0.656	119.4 [7]	0.655	+0.1
Glucitol (sorbitol)	182.2	119.5	0.656	119.5 [7]	0.656	0
Glucuronic acid	194.2	107.7	0.555	110.2 [7]	0.567	−2.3
Glucuronic acid, Na	216.2	92.9	0.430	95.7 [7]	0.443	−2.9
27) Lipids and Constituents:[c]						
Monoacyl phospholipids:						
LysoC$_{16}$PC [20 °C]	495.6	462.4	0.933	463.4	0.935 [19]	−0.2
Diacyl phospholipids:						
diC$_6$PC	453.5	400.5	0.883	392.3	0.865 [19]	+2.1
diC$_7$PC	481.6	432.7	0.898	427.7	0.888 [19]	+1.2
diC$_{14}$PC [23 °C]	677.9	658.1	0.971	652.8	0.963 [9]	+0.8
Steroids:						
Cholesterol [20 °C]	386.7	394.2	1.019	367.0	0.949 [19]	+7.4
[below cmc, in benzene, temperature not specified]	386.7	394.2	1.019	394.8	1.021 [9]	−0.2
Digitonin [20 °C]	1229.5	862.5	0.702	907.4	0.738 [19]	−4.9
Acylcarnitines:						
C$_{12}$Carn	343.5	330.8	0.963	333.2	0.970 [19]	−0.7
C$_{14}$Carn	371.6	363.0	0.977	369.7	0.995 [19]	−1.8
C$_{16}$Carn	399.6	395.2	0.989	400.4	1.002 [19]	−1.3

Table 5. (Continued)

Molecules[a]	Calculated Volumes			Experimental Volumes[b]		Δ
	M (g/mol)	$\bar{V}_c$ (cm^3/mol)	$\bar{v}_c$ (cm^3/g)	$\bar{V}_{exp}$ (cm^3/mol)	$\bar{v}_{exp}$ (cm^3/g)	$\Delta \bar{V}$ (%)
Monoglycerides						
C$_{10}$Glyc	246.4	235.4	0.956	229.1	0.93 [19]	+2.7
C$_{12}$Glyc [45°C]	274.4	267.6	0.975	263.4	0.96 [19]	+1.6
28) Detergents:[c]						
Ionic Detergents:						
Alkyl sulfates:						
NaC$_{12}$SO$_4$ (SDS)	288.4	229.4	0.795	247.6	0.859 [19]	−7.4
[below cmc]	288.4	229.4	0.795	235.0	0.815 [9]	−2.4
Alkyl sulfonates:						
NaC$_{12}$SO$_3$ [31.5 °C]	272.4	223.9	0.822	238.6	0.876 [19]	−6.2
Trimethylammoniumchlorides:						
C$_{12}$TMACl [23 °C]	263.9	280.9	1.064	289.0	1.095 [19]	−2.8
Trimethylammoniumbromides:						
C$_{12}$TMABr	308.4	287.8	0.933	295.4	0.958 [19]	−2.6
Dimethylalkylammoniopropane sulfonates:						
DC$_{12}$APS (SB 12) [30 °C]	335.6	321.0	0.957	321.1	0.957 [19]	0
Nonionic Detergents:						
Polyoxyethylene alkylphenols:						
C$_8\phi$E$_{10}$ (Triton X-100)	646.9	590.4	0.913	591.9	0.915 [56]	−0.3
Polyoxyethylene alcohols:						
C$_{16}$E$_{20}$ (Brij-58)	1123.5	1032.5	0.919	1032.5	0.919 [19]	0
Polyoxyethylene monoacyl sorbitans:						
C$_{12}$SorbE$_{20}$ (Tween-20)	1227.6	1067.3	0.869	1067	0.869 [19]	0
Dimethylalkyl amine oxides:						
DC$_{10}$AO [30 °C]	201.4	222.4	1.104	222.7	1.106 [19]	−0.1
Dimethylalkyl phosphine oxides:						
DC$_{10}$PO [30 °C]	218.3	248.9	1.140	244.1	1.118 [19]	+2.0
Alkylsulfinyl alkanols:						
C$_6$SOC$_2$OH	178.3	170.7	0.957	169.7	0.952 [19]	+0.6
Alkylglycosides:						
C$_8$GS [temperature not specified]	292.4	246.1	0.842	251.2	0.859 [20]	−2.0
29) Biopolymers:[d]						
Polyaminoacids:						
Poly-DL-phenylalanine						
[in toluene, 20 °C]	147.2	116.4	0.791	118.5	0.805 [9]	−1.8
Poly-L-tyrosine						
[in dimethylformamide, 20 °C]	163.2	118.7	0.727	117.5	0.720 [9]	+1.0
Poly-L-glutamic acid						
[in dimethylformamide, 20 °C]	129.1	84.6	0.655	87.1	0.675 [9]	−2.9
Poly-L-lysine						
[in 10% PEG 1000, 20 °C]	128.2	106.1	0.828	101.3	0.790 [9]	+4.7
Polysaccharides:						
Starch	162.1	98.3	0.606	97.5 [6]	0.601	+0.8
Amylose	162.1	98.3	0.606	99.0 [6]	0.611	−0.7
Inulin [20 °C]	162.1	100.3	0.619	97.4	0.601 [9]	+3.0
Hyaluronic acid [temperature not specified]	379.3	232.2	0.612	250.4	0.66 [9]	−7.3
Hyaluronic acid, Na	401.3	217.4	0.542	223.1	0.556 [9]	−2.6
30) Synthetic Polymers:[d]						
Polyethylene glycol	44.1	37.7	0.856	36.9 [6]	0.838	+2.2
Polypropylene glycol	58.1	53.8	0.926	53.1 [6]	0.914	+1.3
Polyethylenimine	43.1	37.3	0.866	37.5 [6]	0.871	−0.5

Table 5. (Continued)

Molecules[a])	M (g/mol)	Calculated Volumes		Experimental Volumes[b])		Δ
		$\bar{V}_c$ (cm^3/mol)	$\bar{v}_c$ (cm^3/g)	$\bar{V}_{exp}$ (cm^3/mol)	$\bar{v}_{exp}$ (cm^3/g)	$\Delta\bar{V}$ (%)
Polyvinylalcohol [20 °C]	44.1	34.5	0.783	33.7	0.765 [57]	+2.4
Poly(diallyldimethyl-						
ammonium chloride) [20 °C]	161.7	137.3	0.849	134.0	0.829 [58]	+2.5
Polyacrylic acid	72.1	48.0	0.666	46.2 [6]	0.641	+3.9
Polyacrylamide [20 °C]	71.1	52.7	0.741	54.7	0.769 [57]	−3.6
Poly(N-methyl)acrylamide	85.1	70.8	0.832	66.6 [6]	0.783	+6.3
Polymethylmethacrylate						
[in benzene]	100.1	85.3	0.852	80.8	0.807 [57]	+5.6
[in dioxane]	100.1	85.3	0.852	81.9	0.818 [57]	+4.2
Polyisobutylene						
[in *n*-hexane]	56.1	64.4	1.148	57.5	1.025 [59]	+12.0
[in octane, 21 °C]	56.1	64.4	1.148	60.3	1.075 [57]	+6.8
[in ethyloctanoate, 22 °C]	56.1	64.4	1.148	62.1	1.106 [57]	+3.7
Polystyrene						
[in benzene]	104.2	95.9	0.921	95.6	0.918 [57]	+0.3
[in cyclohexane]	104.2	95.9	0.921	96.8	0.929 [57]	−0.9
Poly-o-bromostyrene						
[in benzene, 17.5 °C]	183.1	112.5	0.615	109.3	0.597 [60]	+2.9
Poly-m-bromostyrene						
[in benzene, 21 °C]	183.1	112.5	0.615	110.7	0.605 [60]	+1.6

[a]) The designations Na and Na$_2$ indicate the mono- and disodium salt of the corresponding compound, NO$_3$ the nitrate, and Cl and Cl$_2$ the mono- and dichloride, respectively. Sometimes the trivial names of the molecules are given additionally in parentheses. Special conditions of solvent and temperature are listed in brackets. In these cases no attempt has been made to apply corrections to the experimental values; the calculated values, however, refer to water and 25 °C.
[b]) The values for $\bar{V}_{exp}$ or $\bar{v}_{exp}$ were taken from the literature; in general, mean values were used. For detailed references the reader is referred to the original literature cited in the references given.
[c]) Unless otherwise stated, the experimental values are valid for the micellar or aggregated state.
[d]) The molecular weight is given for the monomeric unit.

Table 1 are only operational quantities which have no real physical meaning *per se*. Nevertheless, they enable reliable predictions of volume quantities of physical relevance.

Table 5 not only systematically documents the validity of our approach, but may also serve as an efficient database for various purposes. For example, it is a necessary prerequisite for estimating the accuracy of future volume calculations when experimental reference data are lacking: In such cases the accuracy may be estimated using the tabulated values ($\Delta\bar{V}$) for chemically similar compounds. The table can also facilitate the calculation of volumes of more complex compounds from smaller molecular units (cf. procedure 3 outlined in the legend to Fig. 1).

The partial specific volumes, $\bar{v}$, listed in Table 5 may be used for estimating the volume effects of ligands bound to macromolecules (both specific and unspecific binding); this is a familiar problem in ultracentrifugation. The tabulated $\bar{v}$-values allow clear predictions of the influence of substituents and the behavior of substances in homologous series. In the case of homologs, interpolations may be used for accurate volume estimates.

A promising aspect of our approach is the possibility to derive statements on various kinds of interactions by detailed comparisons of calculated and experimental volume data of homologous substances. In this context the dependence of volumetric effects on the chain length of various compounds, or the different behavior of detergents below and above the critical micellar concentration (CMC) can be addressed.

For the calculation of non-ionic organic compounds, our approach is based on a small number

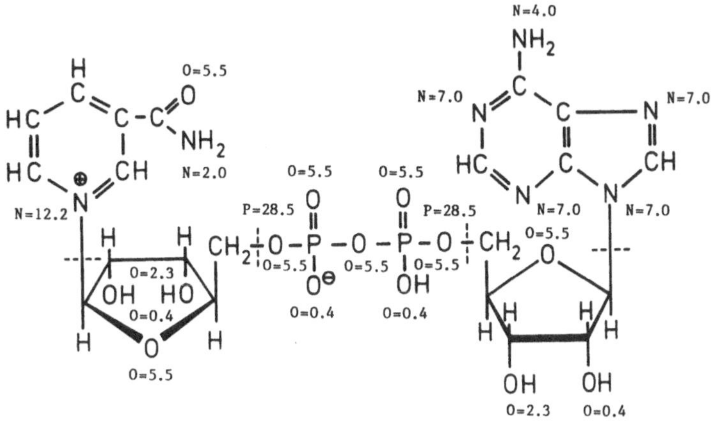

Fig. 1. Calculation of the partial molar volume of nicotinamide adenine dinucleotide (NAD$^+$) from volume increments.

The calculation may be performed using the values in Tables 1–5. For convenience, the data from these tables may be mixed. Some calculation examples are given below. Procedure 1 uses directly the volume increments for atoms and ions given in Tables 1–3. For procedure 2 additionally the volume increments for building blocks (groups) presented in Table 4 are applied. Procedure 3 employs already calculated partial molar volumes of organic and biochemical model compounds (Table 5) and volume increments for atoms, ions and groups for the remaining parts (Tables 1–4). Note: When using calculation procedures 2 or 3, one has to consider some further aspects: i) some atoms or groups may be added or split off upon formation of large molecules; ii) the value for a definite increment may change (e.g., the volumes for O and N); iii) the neighborhood of groups may change (e.g., a first OH may be transformed to a second OH); iv) the charge of some atoms/groups may change; v) calculated volumes of model compounds (Table 5) already contain the contributions for covolume, ring formation and ionization.

Empirical formula for NAD$^+$: $C_{21}H_{27}N_7O_{14}P_2$.
Molar mass for NAD$^+$: 663.44 g/mol.

Procedure 1: Calculation from volume increments for atoms; the values used for O, N, and P are indicated in the figure.

$\bar{V}_c = 21 \times 9.9$ (C) $+ 27 \times 3.1$ (H) $+ 8 \times 5.5$ (O) $+ 2 \times 2.3$ (O) $+ 4 \times 0.4$ (O) $+ 4 \times 7.0$ (N) $+ 4.0$ (N) $+ 12.2$ (N) $+ 2.0$ (N)
$\qquad + 2 \times 28.5$ (P) $+ 12.4$ (V_{CV}) $- 2 \times 8.1$ (V_{RF}) $- 3 \times 6.1$ (V_{RF}) $- 6.8$ ($V_{ES}+$) $- 6.7$ ($V_{ES}-$) $= 409.4$ cm^3/mol.

$\bar{v}_c = \bar{V}_c/M$
$\quad = 0.617$ cm^3/g.

Procedure 2: Calculation from the building blocks indicated in the figure by broken lines.

$\bar{V}_c = 2 \times 76.4$ (ribose residue) $+ 81.7$ (diphosphate residue) $+ 82.8$ (nicotinamide residue) $+ 79.7$ (adenine residue)
$\qquad + 12.4$ (V_{CV}) $= 409.4$ cm^3/mol.

$V_{\text{ribose residue}} = 4 \times 13.0$ (CH) $+ 16.1$ (CH$_2$) $+ 5.5$ (O) $+ 5.4$ (1st OH) $+ 3.5$ (2nd OH) $- 6.1$ (V_{RF}) $= 76.4$ cm^3/mol.

$V_{\text{diphosphate residue}} = 45.4 \left(\begin{smallmatrix} O \\ \| \\ -O-P-O- \\ | \\ O^- \end{smallmatrix} \right) + 28.5$ (P) $+ 2 \times 5.5$ (O) $+ 3.5$ (2nd OH) $- 6.7$ ($V_{ES}-$) $= 81.7$ cm^3/mol.

$V_{\text{nicotinamide residue}} = 71.8$ (phenylene) $- 9.9$ (C) $+ 12.2$ (1st ammonium N) $+ 23.6 \left(\begin{smallmatrix} O \\ \| \\ -C-NH_2 \end{smallmatrix} \right) - 8.1$ (V_{RF})
$\qquad - 6.8$ ($V_{ES}+$) $= 82.8$ cm^3/mol.

$V_{\text{adenine residue}} = 2 \times 13.0$ (CH) $+ 3 \times 9.9$ (C) $+ 4 \times 7.0$ (N) $+ 10.2$ (NH$_2$) $- 8.1$ (V_{RF}) $- 6.1$ (V_{RF}) $= 79.7$ cm^3/mol.

Procedure 3: Calculation from the model compounds presented in Table 5.

$\bar{V}_c = 409.4$ cm^3/mol (calculation in analogy to procedure 2).

$V_{\text{ribose residue}} = 95.8$ (ribose) $- 5.4$ (1st OH) $- 3.5$ (2nd OH) $+ 1.9$ (ΔO) $- 12.4$ (V_{CV}) $= 76.4$ cm^3/mol.

$V_{\text{nicotinamide residue}} = 76.3$ (pyridine) $- 3.1$ (H) $+ 23.6 \left(\begin{smallmatrix} O \\ \| \\ -C-NH_2 \end{smallmatrix} \right) + 5.2$ (ΔN) $- 12.4$ (V_{CV}) $- 6.8$ ($V_{ES}+$) $= 82.8$ cm^3/mol.

$V_{\text{adenine residue}} = 88.1$ (purine) $- 2 \times 3.1$ (H) $+ 10.2$ (NH$_2$) $- 12.4$ (V_{CV}) $= 79.7$ cm^3/mol.

(19) of volume increments for atoms and a few further special increments/decrements for co-volume and ring formation (Table 1). In the case of ionic organic compounds, two additional decrements for ionization (Table 1) and tabulated values for inorganic ions (Table 3) are required. This is already sufficient for calculating an enormous number of organic molecules. For some special problems (e.g., volumes of metalloproteins) some further volume data are needed (Table 2).

The simplicity of our approach is in marked contrast to the rather complicated calculation scheme proposed by Cabani et al. [6] which needs about 90 different group increments (comprising only C, H, N, O, S, F, Cl, Br, I). Moreover, their scheme is not universally applicable to all sorts of organic molecules, since some group increments are missing. The most serious deficiency of their approach, however, is the impossibility to calculate ionic compounds.

Though our approach turned out to be very efficient, it may be improved by future work. Inclusion of additional compounds of the classes already covered by this study could result in a refinement of already existing increments. Inclusion of additional classes of compounds could result in further increments. The latter classes should comprise, for example, unsaturated compounds with isolated, conjugated, or cumulated double or triple bonds, heterocycles containing atoms other than O and N, isomeric molecules (e.g., cis-trans isomerism, ortho, meta, para positions of substituents), further model substances of biochemical relevance. All future improvements, of course, necessitate the precise experimental determination of the partial volumes of the discussed substances in aqueous solution. To some extent the performance of such experiments, however, will be limited by the restricted solubility of many organic compounds in aqueous medium.

Acknowledgements

The authors are much obliged to Prof. D. Schubert, Frankfurt, for his encouraging proposal to perform this study, to Prof. D. Hace, Zagreb, for bringing an essential reference to our attention, and to Prof. R. Jaenicke, Regensburg, and Prof. J. Schurz, Graz, for their kind interest in this work.

References

1. Kopp H (1839) Poggendorff's Ann 47:133–153
2. Kopp H (1889) Liebig's Ann Chem 250:1–117
3. Traube J (1896) Liebig's Ann Chem 290:43–122
4. Traube J (1899) Samml chem chem-tech Vortr 4:255–332
5. Millero FJ, Lo Surdo A, Shin C (1978) J Phys Chem 82:784–792
6. Cabani S, Gianni P, Mollica V, Lepori L (1981) J Solution Chem 10:563–595
7. Høiland H (1986) In: Hinz H-J (ed) Thermodynamic Data for Biochemistry and Biotechnology. Springer-Verlag, Berlin – Heidelberg – New York – Tokyo, pp 17–44
8. Durchschlag H, Zipper P (1993) In: Biškup B, Despotović R, Nemet Z (eds) Proceedings of the 1st Symposium of Croatian Society for Surfactants, Rovinj 1993. Croatian Society for Surfactants, Zagreb, pp 149–163
9. Durchschlag H (1986) In: Hinz H-J (ed) Thermodynamic Data for Biochemistry and Biotechnology. Springer-Verlag, Berlin – Heidelberg – New York – Tokyo, pp 45–128
10. Millero FJ (1971) Chem Rev 71:147–176
11. Durchschlag H (1989) Colloid Polym Sci 267:1139–1150
12. Schmidt GC (1890) Monatsh Chem 11:51–57
13. Cohn EJ, Edsall JT, eds (1943) Proteins, Amino Acids and Peptides as Ions and Dipolar Ions. Reinhold, New York. (1965) Reprint by Hafner, New York
14. Zamyatnin AA (1972) Progr Biophys Mol Biol 24:107–123
15. Zamyatnin AA (1984) Annu Rev Biophys Bioeng 13:145–165
16. Perkins SJ (1986) Eur J Biochem 157:169–180
17. Langridge R, Marvin DA, Seeds WE, Wilson HR, Hooper CW, Wilkins MHF, Hamilton LD (1960) J Mol Biol 2:38–64
18. Van Krevelen DW (1990) Properties of Polymers, 3rd ed. Elsevier, Amsterdam – London – New York – Tokyo, pp 71–107
19. Steele JCH Jr, Tanford C, Reynolds JA (1978) Methods Enzymol 48:11–23
20. Reynolds JA, McCaslin DR (1985) Methods Enzymol 117:41–53
21. Edward JT (1970) J Chem Educ 47:261–270
22. Edward JT, Farrell PG (1975) Can J Chem 53:2965–2970
23. Terasawa S, Itsuki H, Arakawa S (1975) J Phys Chem 79:2345–2351
24. Shahidi F, Farrell PG, Edward JT (1976) J Solution Chem 5:807–816
25. Edward JT, Farrell PG, Shahidi F (1977) J Chem Soc Faraday Trans I 73:705–714
26. Shahidi F, Farrell PG, Edward JT (1977) J Chem Soc Faraday Trans I 73:715–721
27. Shahidi F, Farrell PG (1978) J Chem Soc Faraday Trans I 74:858–868
28. Chothia C (1975) Nature (London) 254:304–308
29. Cabani S, Conti G, Lepori L (1972) J Phys Chem 76:1338–1343
30. Cabani S, Conti G, Lepori L, Leva G (1972) J Phys Chem 76:1343–1347
31. Cabani S, Conti G, Lepori L (1974) J Phys Chem 78:1030–1034
32. Jolicoeur C, Boileau J, Bazinet S, Picker P (1975) Can J Chem 53:716–722

33. Jolicoeur C, Lacroix G (1976) Can J Chem 54:624–631
34. Jolicoeur C, Boileau J (1978) Can J Chem 56:2707–2713
35. Roux G, Perron G, Desnoyers JE (1978) Can J Chem 56:2808–2814
36. Edward JT, Farrell PG, Shahidi F (1979) Can J Chem 57:2887–2891
37. Edward JT, Farrell PG, Shahidi F (1979) Can J Chem 57:2892–2894
38. Zana R (1980) J Polym Sci Polym Phys Ed 18:121–126
39. Letellier P, Gaboriaud R (1981) J Chim Phys 78:829–836
40. Rao MVR, Atreyi M, Rajeswari MR (1984) J Chem Soc Faraday Trans I 80:2027–2032
41. Rao MVR, Atreyi M, Rajeswari MR (1984) J Phys Chem 88:3129–3131
42. Yoshimura Y, Osugi J, Nakahara M (1985) Ber Bunsenges Phys Chem 89:25–31
43. Yoshimura Y, Nakahara M (1985) Ber Bunsenges Phys Chem 89:426–432
44. Yoshimura Y, Nakahara M (1985) Ber Bunsenges Phys Chem 89:1004–1008
45. Yoshimura Y, Nakahara M (1986) Ber Bunsenges Phys Chem 90:58–65
46. Nishimura N, Tanaka T, Motoyama T (1987) Can J Chem 65:2248–2253
47. Makhatadze GI, Medvedkin VN, Privalov PL (1990) Biopolymers 30:1001–1010
48. Kiyosawa K (1991) Biochim Biophys Acta 1064:251–255
49. Noyes RM (1964) J Amer Chem Soc 86:971–979
50. Millero FJ (1972) In: Horne RA (ed) Water and Aqueous Solutions. Wiley-Interscience, New York – London – Sydney – Toronto, pp 519–564
51. Friedman HL, Krishnan CV (1973) In: Franks F (ed) Water, A Comprehensive Treatise, Vol 3. Plenum Press, New York – London, pp 1–118
52. Edsall JT (1953) In: Neurath H, Bailey K (eds) The Proteins, Vol 1, Part B. Academic Press, New York, pp 549–726
53. Lax E, Synowietz C, eds (1967) D'Ans-Lax, Taschenbuch für Chemiker und Physiker, Vol I, 3rd ed. Springer-Verlag, Berlin – Heidelberg – New York, pp 80–105
54. Millero FJ (1972) In: Horne RA (ed) Water and Aqueous Solutions. Wiley-Interscience, New York – London – Sydney – Toronto, pp 565–595
55. Jolicoeur C, Philip PR, Perron G, Leduc PA, Desnoyers JE (1972) Can J Chem 50:3167–3178
56. Stubičar N, Matejaš J, Zipper P, Wilfing R (1989) In: Mittal KL (ed) Surfactants in Solution, Vol 7. Plenum Press, New York – London, pp 181–195
57. Klärner PEO, Ende HA (1975) In: Brandrup J, Immergut EH (eds) Polymer Handbook, 2nd ed. Wiley-Interscience, New York – London – Sydney – Toronto, pp IV 61–113
58. Wandrey C, Görnitz E (1992) Acta Polymer 43:320–326
59. Lederer K, Klapp H, Zipper P, Wrentschur E, Schurz J (1979) J Polym Sci Polym Chem Ed 17:639–648
60. Durchschlag H, Puchwein G, Kratky O, Breitenbach JW, Olaj OF (1970) In: Overberger CG, Fox TG (eds) Polymers and Polymerization (J Polym Sci C 31). Wiley-Interscience, New York, pp 311–343

Received October 28, 1993;
accepted December 6, 1993

Authors' address:

Dr. Helmut Durchschlag
Institut für Biophysik und Physikalische Biochemie
Universität Regensburg
Universitätsstraße 31
93040 Regensburg, FRG

Progress in Colloid & Polymer Science   Progr Colloid Polym Sci 94: 40–45 (1994)

Analysis of interaction of the small heat shock protein hsp25 with actin by analytical ultracentrifugation

J. Behlke, O. Ristau, and A. Knespel

Max Delbrück Centrum für Molekulare Medizin, Berlin, FRG

Abstract: By using analytical ultracentrifugation interactions of the small heat shock protein hsp25 and G-actin have been analyzed. Association constants of about 10^7 reciprocal molar concentration were estimated from the concentration distribution using a computer program based on the nonlinear least-squares methode. In the presence of an excess of actin up to four molecules of this protein were bound to hsp25 particles.

Key words: Analytical ultracentrifugation – protein assemblies – complex formation – association constants – nonlinear least-squares techniques

Introduction

Cells which are exposed to different kinds of stress (heat, chemicals, etc.) react with the expression of specific proteins called heat shock proteins (hsp) [1]. They are classified according to their molecular mass and subdivided in several groups. One of these, the small hsp with molecular masses of 15–30 kDa shows sequence homology with the lens protein α-crystallin [2–4]. Furthermore, hsp25 and α-crystallin are able to form large sphere-like complexes containing up to 30 and more monomers [5–7]. In which manner such particular structures can help cells to survive stress situations is unknown so far. Recently, Miron et al. [8] described the inhibition of actin polymerization by hsp25 from turkey and chicken smooth muscle. Because hsp25 can form assemblies of different size depending on the ionic strength, we have analyzed such systems and its interaction with actin using the technique of analytical ultracentrifugation. From mixtures of both proteins in different compositions hsp-actin complexes with association constants of about 10^7 reciprocal molar concentration were obtained.

Material and methods

Recombinant murine hsp25 was prepared and purified as given in [7]. The protein concentration was determined spectrophotometrically using an absorption coefficient of $A_{280\,nm}^{1\%} = 10.0\,cm^{-1}$. Actin from rabbit muscle was isolated according to Spudich and Watts [9] and purified by polymerization/depolymerization steps and subsequent centrifugation to remove denatured protein [10]. An absorption coefficient $A_{290\,nm}^{1\%} = 6.3\,cm^{-1}$ was used to determine the actin concentration. The heat shock protein and actin were extensively dialyzed against buffer A (2 mM Tris-HCl, pH 8.0, 0.2 mM Na_2ATP, 0.2 mM $CaCl_2$, 0.5 mM β-mercaptoethanol, 0.005% NaN_3).

Sedimentation-, diffusion- and molecular mass determinations were carried out by an analytical ultracentrifuge Spinco E with Schlieren-, interference- or UV-optics, monochromator and photoelectric scanner, respectively. Sedimentation and diffusion experiments were performed using a double sector synthetic boundary cell as described in [11]. Molecular mass determinations were carried out by sedimentation equilibrium technique with the interference optics and six-channel cells as given in [11]. According to the different distances r from the center of revolution the inner, middle or outer compartments of the cell were filled with 100 μl protein solution of hsp and mixtures of hsp and actin, respectively. All samples contained 0.3–0.45 mg/ml of protein. To avoid a possibly discriminatory influence of gravity on the stability of protein complexes at different positions of r, some experiments with hsp as well as hsp-actin were repeated under equal conditions using the same compartment of the cell. Molecular mass (M) determinations were performed either from the slope $\ln c/r^2$ according to Eq. (1) which is valid for ideal solutions,

$$M = \frac{2RT}{(1 - \rho\bar{v}_i)\omega^2} * \frac{d\ln c}{d(r^2)}, \quad (1)$$

with R being the gas constant, T the absolute temperature, ρ the density of solution, $\bar{v}_i$ the partial specific volume of polymer component i and ω the angular velocity [12], or by direct estimation of M according to Eq. (2) which is obtained from formula (1) by integration, assuming ρ and $\bar{v}_i$ are not a function of r [13]:

$$c_r = c_0 * e^{M*F}. \quad (2)$$

Herein

$$F = [(1 - \rho\bar{v}_i)\omega^2 (r^2 - r_0^2)]/2RT. \quad (3)$$

In these equations c_0 is the concentration at the reference radius r_0. For analyzing the non-linear regression a computer program polymol was used. It fits the different values $c = f(r)$ according to the least squares procedure of Levenberg [14] and Marquardt [15] in the version of Wynne und Wormell [16].

Beside the molar mass also the concentration can be determined for one substance. However, in the case of two reacting components it is useful to estimate the molecular masses of each component from the concentration distributions in separate compartments or experiments under identical conditions and use these data as fixed values in the reacting system. By this procedure the number of free parameters is reduced.

Homopolymers (P) with a supramolecular structure should be able to form more than one binding site for the interaction with other macromolecules which should be considered here as ligands (L). Assuming *equal* binding sites for all

(maximal n) bound ligands from the statistical point of view the following relations between the binding constants have to be considered.

According to the equilibrium:

$$K_i = \frac{[PL_i]}{[PL_{i-1}]*[L]} \quad (4)$$

the stepwise binding constant of the i-th ligand can be written as

$$K_i = \left(\frac{n+1-i}{n*i}\right)*K_1, \quad (5)$$

where $i = 1, n$.

The expression for the overall binding constant between monomeric components (here the homopolymer P should be considered as a monomeric particle)

$$\beta_j = \frac{[PL_j]}{[P]*[L]^j} \quad (6)$$

can be modified because of the validity

$$\beta_j = K_1 * K_2 * \cdots K_j, \quad (7)$$

and written in the following way [17]

$$\beta_j = (K_1)^j * \frac{1}{n^j}\binom{n}{j}, \quad (8)$$

where $j = 1, n$.

The model function for the concentration distribution in sedimentation equilibrium experiments can be formulated by means of the modified binomial coefficients (Eq. (8)) without considering the virial coefficients in the following way:

$$c_r = c_{0P}*e^{M_P*F} + c_{0L}*e^{M_L*F}$$
$$+ c_{0P}*\sum\frac{1}{n^j}\binom{n}{j}*(c_{0L}*K_1)^j*e^{(M_P + j*M_L)F} \quad (9)$$

Equation (9) was fitted to the experimental data $c = f(r)$ by the least squares method. With respect to the limited exactness of the experimental data and the marked mutual dependence of the parameters in this model it is important to reduce its number as far as possible. Since the molecular masses can be determined in parallel experiments, we have to estimate only the concentrations c_{0P} and c_{0L} as well as the equilibrium constant. A further reduction of the number of parameters is attainable considering the mass balance as constraint. The total concentration (c^t) of the two components P and L can be obtained by dissolving Eqs. (10, 11):

$$c_P^t = \frac{c_{0P}}{(r_b - r_m)} \int_{r_m}^{r_b} e^{M_P * F} dr + \frac{c_{0P}}{(r_b - r_m)}$$

$$\times \int_{r_m}^{r_b} \sum_j \frac{M_P}{M_P + j*M_L} (c_{0L}*K_1)^j$$

$$*e^{(M_P + j*M_L)*F} dr \qquad (10)$$

$$c_L^t = \frac{c_{0L}}{(r_b - r_m)} \int_{r_m}^{r_b} e^{M_L * F} dr + \frac{c_{0P}}{(r_b - r_m)}$$

$$\times \int_{r_m}^{r_b} \sum_j \frac{j*M_L}{M_P + j*M_L} (c_{0L}*K_1)^j$$

$$*e^{(M_P + j*M_L)*F} dr \ , \qquad (11)$$

with r_b and r_m being the radius at the bottom or meniscus of the cell, respectively. The two numeric integration steps have to be carried out in each iteration cycle and dissolved iteratively for c_{0L}. and c_{0P}. At least only the equilibrium constant remains to be estimated. For this procedure the fringes displacement must be calibrated for true concentrations. The maximal number of binding sites cannot be estimated.

Results

1. Sedimentation velocity experiments

Globular or G-actin in buffer A is characterized by sedimentation coefficient of 3.2 (S) and diffusion coefficient of $6.7*10^{-7}$ cm²/s. From these data a molecular mass of about 42 kDa can be calculated, indicating actin is free of aggregates.

Recombinant hsp25 in buffer I (20 mM Tris-HCl, pH 7.6, 10 mM MgCl₂, 30 mM NH₄Cl, 0.5 mM dithioerythritol, 0.05 mM NaN₃, 2 μM phenylmethanesulfonyl fluoride) sediments with 18–20 (S) and is characterized by diffusion coefficients of $2.6 - 3.0*10^{-7}$ cm²/s. Molecular masses of 600 kDa or more result from these values, indicating about 30 molecules are assembled in the particles.

Because the high concentration of Mg⁺⁺ in the buffer leads to a polymerization of G-actin, hsp25 was analyzed also in buffer A with lower ionic strength. Under these conditions sedimentation coefficients of the stress protein are reduced to

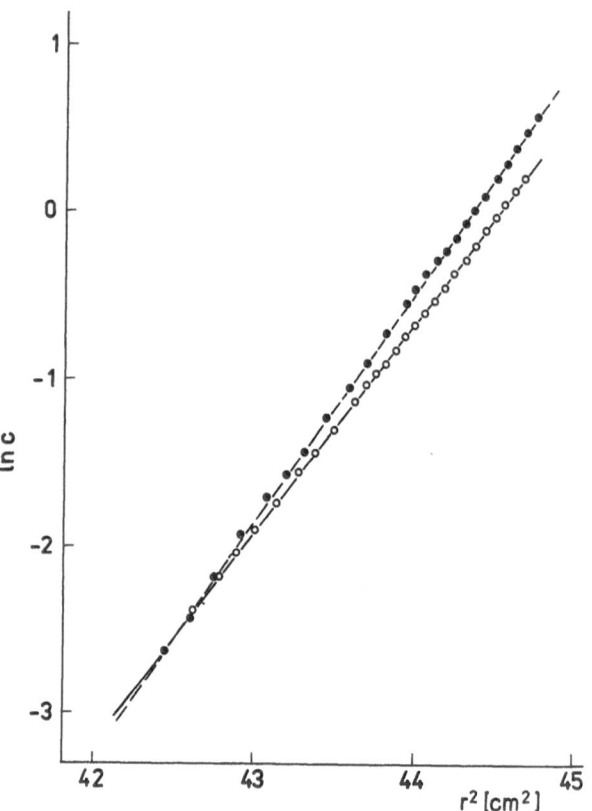

Fig. 1. Ln c/r^2 plots of the concentration distribution of hsp25 (○) and hsp25 + G-actin (●). Initial concentration: 0.42 mg/ml hsp25, 0.033 mg/ml actin dissolved in buffer A. Equilibrium speed 8000 rpm, temperature 14 °C

about 13 (S), corresponding to molecular masses of 300–350 kDa or particles of 14–15 molecules.

2. Sedimentation equilibrium experiments

Hsp samples and some mixtures of hsp with actin extensively dialyzed against buffer at 4 °C were analyzed by sedimentation equilibrium runs. Figure 1 demonstrates ln c/r^2 plots of the concentration distribution of hsp25 and hsp25-actin. From the slope a molecular mass of 331 kDa was calculated for the stress protein dissolved in buffer A. On the average, these particles contain 14–15 molecules. The ln c/r^2 plot obtained for a mixture of 0.42 mg/ml or $1.25*10^{-6}$M hsp and 0.033 mg/ml or $8.1 \cdot 10^{-7}$M G-actin results also in a straight line. Using Eq. (1) from the slope $d\ln c/d(r^2)$ an average molecular mass of 364 kDa was calculated. The difference between the two values amounts to 33 kDa, less than the molecular mass

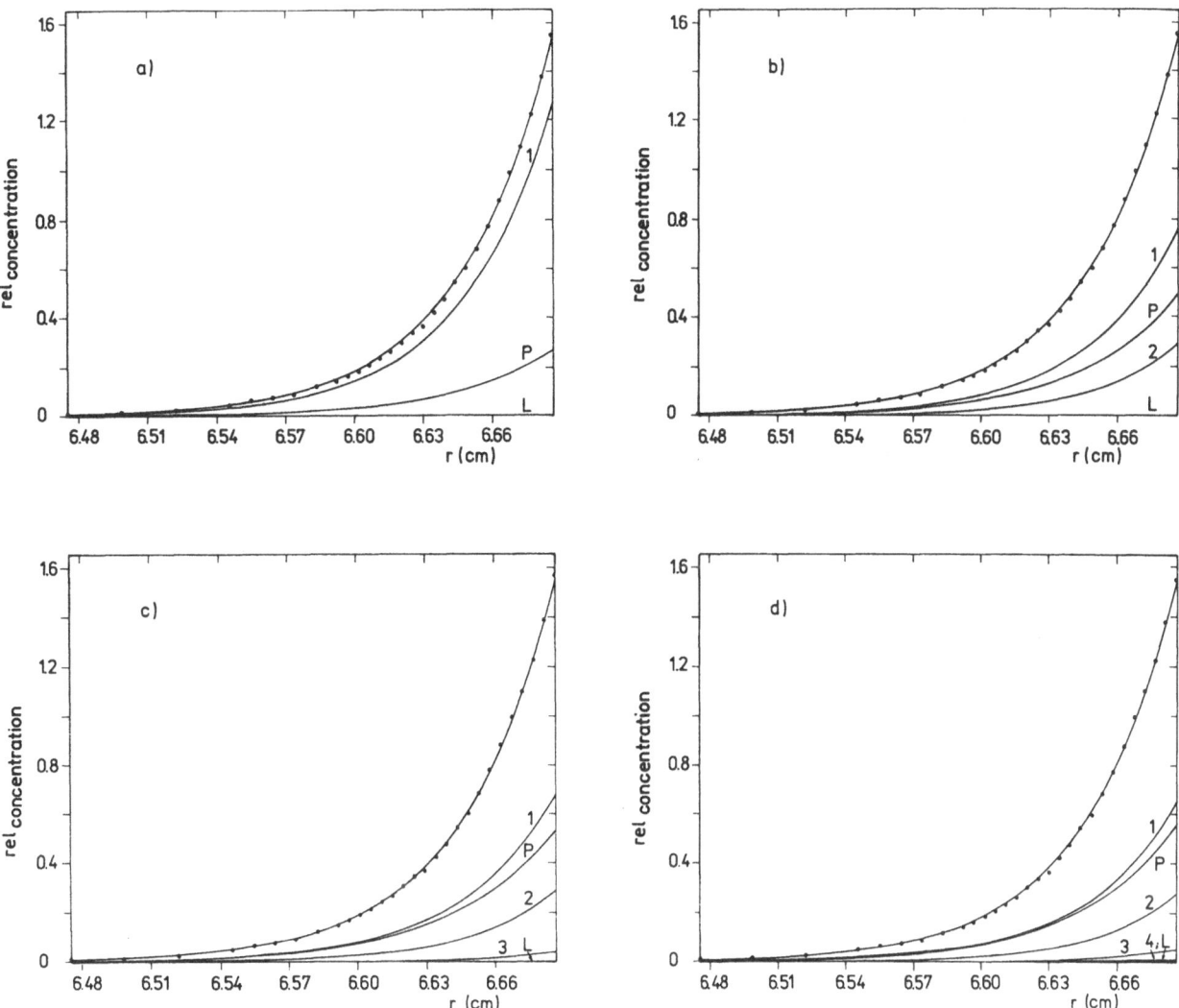

Fig. 2. Calculated concentration distributions $c = f(r)$ of hsp25 (P), actin (L) as well as their different complexes assuming binding sites for one (a), two (b), three (c) or four (d) ligands, respectively. The numbers mean complexes with one, two, three or four bound ligands. Experimental data (–•–•–•–) are given in comparison. Conditions as mentioned in Fig. 1

of actin, which amounts to 42 kDa. Taking into account an excess of hsp particles, we can expect a partial complex formation between hsp and actin beside of unligated stress protein and presumably no or only few free G-actin. This behavior corresponds to the rather straight $\ln c/r^2$ plot which reflects only small differences in the molecular mass.

To analyze the complex formation in more detail, the experiments were additionally analyzed by Eq. (9) using the $c = f(r)$ data. Although the above-mentioned experiment contains less actin, which has to be considered as ligand, we cannot exclude the binding of more than one molecule to some hsp particles. Figure 2 represents the experimental and theoretical radial concentration distributions assuming one, two, three or four possible binding sites, respectively. The partial concentrations of the species were calculated considering the mass balance of both proteins in the experiment, and the size of hsp obtained separately under identical conditions. As expected from the straight line $\ln c/r^2$ plot (Fig. 1), most of G-actin added to the hsp particles are bound. Assuming a 1:1 stoichiometry for the complex formation an association constant of about 10^7

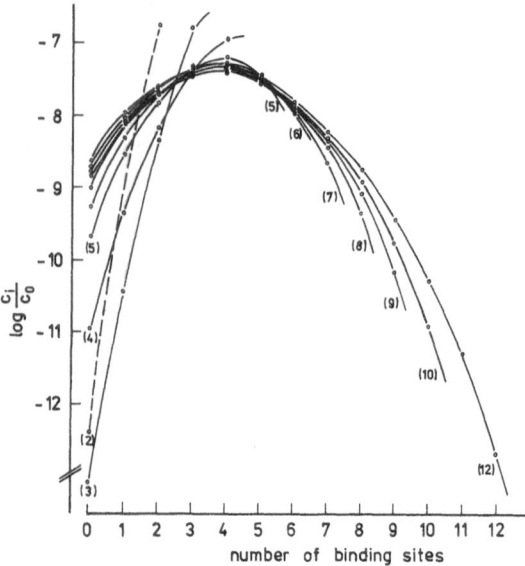

Fig. 3. Calculated partial concentrations of a sedimentation equilibrium experiment with $1.68*10^{-6}$ M hsp25 and $7.14*10^{-6}$ M G-actin in buffer A assuming different numbers of binding sites. The numbers in parentheses mean maximal binding sites. Conditions as given in Fig. 1

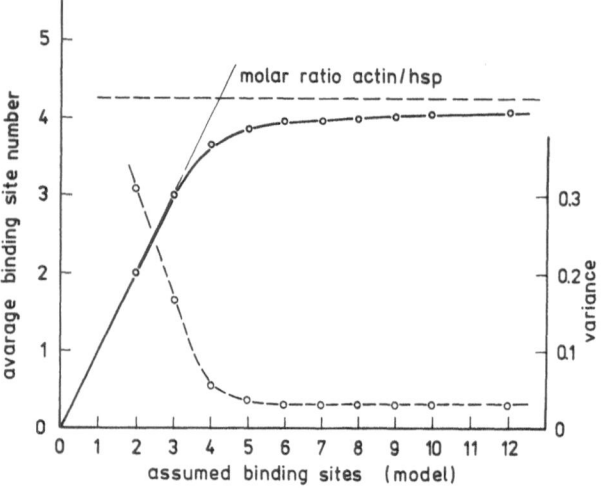

Fig. 4. Calculated values for the average ligand binding and the variance (dotted line) assuming different numbers of binding sites. Experimental conditions as given in Fig. 3

reciprocal molar concentration can be estimated. When assuming also an isomerization or formation of complexes with more ligands, necessarily the amount of free stress protein (P) increases in comparison with the data obtained for a 1:1 stoichiometry (Fig. 2). To get more information about the maximal binding site number of hsp25 particles experiments with a molar excess of actin were carried out and fitted to different models. As expected, the analysis of such distribution curves does not allow a good fit assuming only 1:1 complexes. Since the hsp particles are only roughly of the same size the determination of the maximal binding site number is difficult. However, the average number of bound ligands can be estimated from the calculated partial concentration distribution curves by using different models. Figure 3 represents some distribution curves for a sedimentation equilibrium experiment with $1.68*10^{-6}$ mol hsp and $7.14*10^{-6}$ mol G-actin. Complexes with four bound ligands were observed in the highest concentration. This result is in agreement with the molar ratio actin/hsp of 4.25 in this experiment. The average number of binding sites can be determined also when plotting the calculated amount of bound ligands against the number of assumed binding sites

(Fig. 4). The obtained curve looks like a titration curve from which the most probably average binding place number 4 can be derived. Furthermore, the data obtained for the variance decrease up to attaining the most probable model and then they remain constant.

Discussion

Numerous catalytic reactions take place in cellular systems. As a prerequisite to understanding such behavior in more detail, information about the kind of interactions between the components, its stoichiometry, etc. is of interest. Methods which yield quantitative results in the relations between different macromolecules are rather scarce. For analyzing self-associating systems analytical ultracentrifugation is a powerful technique. Here, we have tried to also analyze a heterologous system using this method. The least-squares procedure, which is useful to analyze data of analytical ultracentrifugation [13], can be taken also when considering some constraints to reduce the number of parameters which have to be estimated. The reduction of free parameters give us the chance to estimate binding constants as an important parameter of interacting systems. But we must consider that constraints in this model contain some oversimplifications such as strictly equal binding sites and the assumption of conservation of molecular masses in components of the interacting systems. Further work will show whether the chosen constraints are justified.

The obtained association constant of about 10^7 reciprocal molar concentration for the reaction between hsp25 and actin is moderate. The surface of such sphere-like hsp particles is large enough to bind 7–8 molecules of G-actin. However, by this method no information about preferential binding of actin can be given.

For future work it is of interest to analyze the interaction between different kinds of hsp25 (phosphorylated and unphosphorylated) and actin for understanding the maturity of the biological reaction in the cell.

Acknowledgements

Thanks are extended to Dr. M. Gaestel for preparing hsp25 and Mrs. B. Bödner for technical assistance.
This work was supported by the Deutsche Forschungsgemeinschaft (Be 1404/1-1).

References

1. Lindquist S (1986) Ann Rev Biochem 55:1151–1191
2. Ingolia TD, Craig EA (1982) Proc Natl Acad Sci USA 70:2360–2364
3. Hickey E, Brandon SE, Potter R, Stein G, Stein J, Weber LA (1986) Nucleic Acid Res 14:4127–4145
4. DeJong WW, Leunissen JAM, Leenen PJM, Zweers A, Versteeg M (1988) J Biol Chem 263:5141–5149
5. Siezen RJ, Bindels JG, Hoenders HJ (1980) Eur J Biochem 111:435–444
6. Arrigo AP, Welch WJ (1987) J Biol Chem 262:15359–15369
7. Behlke J, Lutsch G, Gaestel M, Bielka H (1991) FEBS Lett 288:119–122
8. Miron T, Vancompernolle K, Vanderkerckhove J, Wilchek M, Geiger B (1991) J Cell Biol 114:255–261
9. Spudich JA, Watts S (1971) J Biol Chem 246:4866–4871
10. Pardel JD, Spudich JA (1982) Methods Enzymol 85:164–181
11. Behlke J, Knespel A, Glaser RW, Pleißner K-P (1991) Progr Colloid Polym Sci 86:30–35
12. Yphantis DA, Waugh DF (1956) J Phys Chem 60:630–636
13. Johnson LM, Correia JJ, Yphantis DA, Halvorson HR (1981) Biophys J 36:575–588
14. Levenberg K (1944) Quart Appl Math 2:164–172
15. Marquardt J (1963) Soc Indust Appl Math 11:431–441
16. Wynne CG, Wormell PMJH (1963) Applied Optics 2:1233–1238
17. Wyman J, Gill SJ (1990) Binding and Linkage, University Science Books Mill Valley, USA, 1990

Received June 11, 1993;
accepted October 25, 1993

Authors' address:

Prof. Dr. J. Behlke
Max Delbrück Centrum für Molekulare Medizin
Robert-Rössle-Straße 10
13121 Berlin

Progress in Colloid & Polymer Science   Progr Colloid Polym Sci 94:46–53 (1994)

Ultracentrifugal analysis of protein-nucleic acid interactions using multi-wavelength scans

M. S. Lewis[1]), R. Shrager[2]), and S. J. Kim[3])

[1]) Biomedical Engineering and Instrumentation Program, National Center for Research Resources
[2]) Physical Sciences Laboratory, Division of Computer Research and Technology
[3]) Laboratory of Biochemistry, National Cancer Institute, National Institutes of Health, Bethesda, Maryland, USA

Abstract: The use of analytical ultracentrifugation for the investigation of the interactions of proteins and nucleic acids has been studied by computer simulations and experimentally in the Beckman XL-A analytical ultracentrifuge. The technique involves obtaining absorbency data at wavelengths in the range of 230 to 246 nm as well as at 260 and 280 nm for the protein solution, the nucleic acid solution, and a solution of these reactants and their complex. The data from the protein and nucleic acid solutions permits calculation of the molar extinction coefficients of these reactants as functions of wavelength which then can be used to construct an extinction coefficient matrix. The data from the solution of reactants and complex permits constructing an absorbency matrix which is a function of radius and wavelength. These matrices may then be used to obtain a data matrix with radial positions in the first column and molar concentrations of protein plus complex and nucleic acid plus complex in the second and third columns. This data matrix can then be analyzed by mathematical modeling to obtain the value of the natural logarithm of the molar equilibrium constant. The computer simulation study demonstrates that the generating value of $\ln K$ is recovered with very little error in spite of the presence of substantial random error added to the absorbency data; the experimental study demonstrates that the method can be applied with equal facility to data obtained with the XL-A ultracentrifuge. The temperature dependence of these values of $\ln K$ can then be used to obtain values of ΔG^0, ΔH^0, ΔS^0 and ΔC_p^0 for the interaction.

Key words: Ultracentrifugal analysis – proteins – nucleic acids – molecular interactions

1. Introduction

The majority of protein-DNA interaction studies have utilized biochemical techniques that include filter-binding assays [1], gel mobility shift [2, 3, 4], and chemical modification or chemical protection and footprinting [5, 6, 7, 8]. These methods and the application of numerical methods to the analysis of data from them have been reviewed recently [9]. Physicochemical approaches have included fluorescence analysis [10, 11], calorimetry [12], and analytical ultracentrifugation [13, 14]. Each of these techniques has distinctive advantages and disadvantages. In particular, analytical ultracentrifugation has the advantages that it is rigorously based upon reversible thermodynamics, and the reactants and product of an interaction each have a uniquely defined concentration gradient that can be resolved by appropriate mathematical analyses to

**) Civilized Software, 7735 Old Georgetown Road, Bethesda, MD 20814 USA give the desired value of ln K, and hence ΔG^0, with a minimum of assumptions.

Prior to the development of the Beckman XL-A analytical ultracentrifuge, protein-nucleic acid interactions investigated in the ultracentrifuge were limited by deficiencies in the absorption optical system intrinsic to the Beckman Model analytical ultracentrifuge such as mediocre monochromaticity, limitation to wavelengths above 260 nm, and limitation to maximum solute absorbencies in the centrifuge cells of about one absorbency unit. The wavelength limitation creates a potential problem with hyper- or hypochromism of the protein-nucleic acid complex; this can limit the accuracy of the results of the analysis unless studies are undertaken to evaluate the extent of change in the absorbency of the complex. Optimal analysis to the data requires a complex analytical approach such as the method of implicit constrains that utilizes conservation of mass within the ultracentrifuge cell to limit values possible for parameters [13, 15].

The new Beckman XL-A analytical ultracentrifuge has significantly enhanced performance with respect to wavelength range (200–800 nm), monochromaticity ($\pm$ 2 nm bandpass), and absorbency range (0–3). The analytical method to be described here utilizes these properties, particularly those of wavelength range and monochromaticity, to minimize the effects of hyper- or hypochromism and to obtain results with enhanced accuracy without the experimentally stringent requirement of conservation of mass in the ultracentrifuge cell.

2. Theory

For simplicity, let us consider only one radial position, r, within the solution column in the ultracentrifuge cell and measure the absorption at a single wavelength, λ. The total molar concentrations of protein, $C_{P,T,r}$, and of nucleic acid, $C_{N,T,r}$ at this radial position are then given by:

$$C_{P,T,r} = C_{P,r} + C_{PN,r} = C_{P,r} + K_{PN} C_{P,r} C_{N,r} \tag{1}$$

$$C_{N,T,r} = C_{N,r} + C_{PN,r} = C_{N,r} + K_{PN} C_{P,r} C_{N,r}, \tag{2}$$

where $C_{P,r}$ is the uncomplexed protein concentration, $C_{N,r}$ is the uncomplexed nucleic acid concentration, $C_{PN,r}$ is the concentration of the complex, and K_{PN} is the equilibrium constant for the formation of the complex.

Since the absorbency of a solution at a particular wavelength is the product of the molar concentration and the extinction coefficient of the solute at that wavelength, the absorbencies of the protein, the nucleic acid, and the complex are given by:

$$A_{P,r,\lambda} = C_{P,r} E_{P,\lambda} \tag{3}$$

$$A_{N,r,\lambda} = C_{N,r} E_{N,\lambda} \tag{4}$$

$$A_{PN,r,\lambda} = C_{PN,r} E_{PN,\lambda} = C_{PN,r}(E_{P,\lambda} + E_{N,\lambda}) \tag{5}$$

Equation (5) explicitly assumes that there is no hyper-or hypochromism at the wavelengths being observed. The optical system of the ultracentrifuge permits only the observation of the total absorption as a function of radial position at a particular wavelength, which is given by:

$$A_{T,r,\lambda} = A_{P,r,\lambda} + A_{N,r,\lambda} + A_{PN,r,\lambda} \tag{6}$$

and substituting Eqs. (3), (4) and (5) in Eq. (6) and rearranging then gives:

$$A_{T,r,\lambda} = E_{P,\lambda} C_{P,T,r} + E_{N,\lambda} C_{N,T,r} \tag{7}$$

Thus, if we have A at wavelengths λ_i, i = 1 to $m > 1$, then Eq. (7) becomes a set of m equations in the two unknowns $C_{P,T,r}$ and $C_{N,T,r}$. The m A's from a row vector **A**, the E's form a 2 x m matrix of coefficients, **E**, and the C's form a row 2-vector, **C**, such that **A** = **CE**. The least-squares solution to this set is **C** = **AE**$^+$, where **E**$^+$ is the pseudo-inverse of **E** [16].

Extending this principle, let **A** be an $n \times m$ matrix of absorbencies such that $A_{i,j}$ is the absorbency for radius r_i and wavelength λ_j. Thus, each column of **A** can be plotted versus r with λ fixed, and each row of **A** can be plotted versus λ with r fixed. In similar fashion, **E** is a $2 \times m$ matrix of molar extinction coefficients, where row 1 is for protein, row 2 is for nucleic acid, and each row can be plotted versus λ. In this context, **C** becomes an $n \times 2$ matrix computed by **C** = **AE**$^+$, where the two columns of **C** contain total molar concentrations, uncomplexed and complexed, of protein and nucleic acid respectively, and each column of **C** can be plotted versus r. We now concatenate **C**, the dependent variables, with a vector of the radial positions, the independent variable, to form a three-column data matrix suitable for analysis by non-linear, least-squares curve-fitting. We use as mathematical models the equations

$$C_{r,P,T} = C_{b,P}\exp(A_P M_P(r^2 - r_b^2))$$

$$+ C_{b,P}C_{b,N}\exp(\ln K$$

$$+ (A_P M_P + A_N M_N)(r^2 - r_b^2)) + \varepsilon_P \quad (8)$$

$$C_{r,N,T} = C_{b,N}\exp(A_N M_N(r^2 - r_b^2))$$

$$+ C_{b,P}C_{b,N}\exp(\ln K + (A_P M_P$$

$$+ A_N M_N)(r^2 - r_b^2)) + \varepsilon_N \quad (9)$$

and jointly fit columns one and two with Eq. (8) and columns one and three with Eq. (9). The value of $\ln K$, the natural logarithm of the molar constant, and the values of $C_{b,P}$ and $C_{b,N}$, the concentrations of uncomplexed protein and nucleic acid at the radius of the cell bottom, r_b, are fitting parameters common (global) to both equations. The values of ε_P and ε_N, the small baseline error terms, are fitting parameters applicable only (local) to Eqs. (8) and (9) respectively. M_P and M_N are the values of the molar mass of the protein and nucleic acid, respectively; A_P and A_N (not to be confused with A, the absorbencies) are defined by:

$$A = (\partial\rho/\partial c)_\mu \omega^2 / 2RT, \quad (10)$$

where ω is the rotor angular velocity, R is the gas constant, and T is the absolute temperature. The thermodynamically more appropriate term $(\partial\rho/\partial c)_\mu$, the derivative of solution density with respect to solute concentration at constant chemical potential, is essentially equal to $(1 - \bar{v}\rho_0)$ at the limit as the solute concentration approaches zero; $\bar{v}$ is the partial specific volume of the appropriate solute and ρ_0 is the solvent density. The chapter by Durchschlag should be consulted for a concise review of this topic [17]. The values of ω, R, T, M_P, and M_N are known; the experimental determination of the values of $(\partial\rho/\partial c)_\mu$ will be described later.

3. Computer simulations

Initial evaluation of a method of analysis such as that described in the previous section is best performed by computer simulation. For this purpose, we have used the mathematical modeling system MLAB** operating on a 80486 computer for both the simulations and the data analysis. Using data from our ultracentrifugal study on the interaction of the enzyme DNA polymerase-β with the synthetic oligonucleotide pd(T)$_{16}$ [13], the C matrix was generated using equimolar concentrations of the reactants, a value of 11.276 for $\ln K$, a rotor speed of 20000 rev min^{-1}, and a temperature of 14°.

Figure 1 illustrates the absorption spectra of equimolar concentrations of the enzyme, of the oligonucleotide and of the mixture of both obtained with a spectrophotometer. Wavelengths from 237 nm to 245 nm were selected so that protein absorbence was dominant at the lower wavelengths and nucleic acid absorbance was dominant at the higher wavelengths. The values of the extinction coefficients at 237, 239, 241, 243,

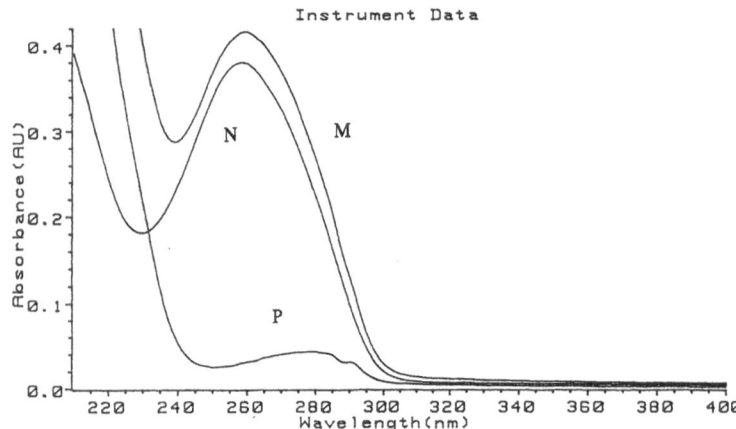

Fig. 1. Absorption spectra of equimolar concentrations of DNA polymerase-β, pd(T)$_{16}$, and an equimolar mixture of both obtained experimentally using a spectrophotometer. The spectra of the protein, the DNA and the mixture are labeled P, N, and M, respectively

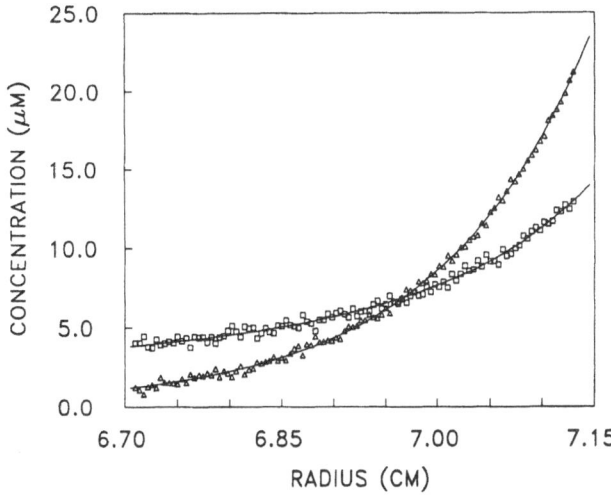

Fig. 2. Distributions of the molar concentrations for the interaction of DNA polymerase-β and the oligonucleotide $pd(T)_{16}$ in a 1 to 1 molar ratio at equilibrium at 20000 rev min^{-1} and $14.0°$. The triangles are for protein plus protein-DNA complex; the squares are for DNA plus protein-DNA complex; the curves are the fitted functions as described in the text. The RMS error is 0.230×10^{-6}M

and 245 nm were obtained by calculating the ratios of the absorbencies at these wavelengths to the absorbency at 260 nm for the oligonucleotide and at 280 nm for the enzyme and then multiplying these ratios by the known extinction coefficients of the oligonucleotide and the enzyme, respectively. The **E** matrix was generated using these values, and the **A** matrix was then obtained as the product of **C** and **E**. The **A** matrix then had normally distributed random error with a mean of zero and a standard error of 0.01 absorbency units added to every element in the matrix. This produced the equivalents of rather noisy scans, about double the preferred noise level and near the upper limit of what would be considered acceptable data from an XL-A ultracentrifuge. Such a noise level represents a stringent test for the method.

A new and noisy **C** matrix was then obtained as the product of the noisy **A** matrix and the pseudoinverse $\mathbf{E}^{+}$; the radius vector was concatenated to this **C** matrix, and the resultant data matrix was fit using Eqs. 8 and 9 as mathematical models. The result of this fit is shown in Fig. 2. In this particular case, a value of 11.266 ± 0.029 was obtained for $\ln K$ with a root-mean-square error for the fit of 0.230×10^{-6} absorbency units. This value of $\ln K$ differs from the generating value of 11.276 by less than 0.09%. A limited Monte Carlo study indicated that such results are typical, and thus the method appears to be quite reliable in terms of returning good values for $\ln K$, at least for data with normally distributed error.

4. Experimental

In order to evaluate the procedure experimentally, we studied the interaction of a heat shock factor peptide with an appropriate DNA in the XL-A analytical ultracentrifuge. The polypeptide encompassed the DNA-binding domain of heat shock factor, spanning the 33 to 163 region of the full-length *Drosophila* HSF, and exhibited purity in excess of 99%. Sequence-specific DNA binding activity was demonstrated by DNAse I footprinting. The molar mass, calculated from the amino acid sequence, was 15147, and the molar extinction coefficient was 9080 mol^{-1} at 280 nm. The synthetically prepared DNA was a 13-base pair double-stranded heat shock element with a $-$GAA$-$ binding site. The compositional molar mass was 8007, and the molar extinction coefficient was 84200 mol^{-1} at 260 nm.

Solutions of peptide, DNA, and a 2:1 molar ratio mixture of peptide and DNA were simultaneously centrifuged to equilibrium at a rotor speed of 14000 rev min^{-1} at a temperature of $4°$. The 2:1 molar ratio was used to increase the fraction of the total absorbance attributable to the peptide at the lower wavelengths. The buffer used was 100 mM KCl, 10 mM K phosphate, pH 6.3, and 1 mM EDTA. Equilibrium was considered to have been attained with the scans at 280 nm and 260 nm had been invariant for 24 h. This was attained by 90 h, and final scans were than taken at 280, at 260, and from 246 through 230 nm at 2 nm increments.

5. Data analysis and results

The first step in the analysis was to obtain the molar extinction coefficients and the values of $(\partial\rho/\partial c)_\mu$ for the peptide and the DNA. The concentration distribution of a homogeneous, ideal solute at ultracentrifugal equilibrium in terms of absorbency as a function of radial position is given by:

$$A_r = A_b \exp\left((M(\partial\rho/\partial c)_\mu/2RT)(r^2 - r_b^2)\right), \quad (11)$$

where the terms have their usual meaning and A_b is the absorbance at the radial position of the cell bottom. Five scans of the peptide solution at wavelengths of 230 nm through 238 nm were jointly fit for the values of the A_b parameters as functions of wavelength and the value of $(\partial\rho/\partial c)_\mu$ as a global fitting parameter. The scans had the greatest absorbency at these wavelengths, and the resultant value of $(\partial\rho/\partial c)_\mu$ was the best attainable. Because of memory space requirements, the overlay version of MLAB used for these analyses could not handle more than 1000 to 1200 data points in simultaneous fitting, and so it was not possible to fit all the data sets simultaneously. The newly released extended memory and Unix versions of MLAB do not have this problem. The scans at 240 through 246 nm and also at 280 nm were then fit for the values of A_b, with the value of $(\partial\rho/\partial c)_\mu$ fixed. A value of 0.2616 ± 0.0045 was obtained for $(\partial\rho/\partial c)_\mu$ of the peptide, with an RMS error of 0.00520 absorbency units for the 5-scan joint fit, which is shown in Fig. 3.

The same procedure was followed for the DNA solution, except that the five scans used for the joint fit to obtain the values of $(\partial\rho/\partial c)_\mu$ and A_b were 238 nm through 246 nm. The scans at 230 nm through 236 nm and also at 260 nm were then fit for the values of A_b, with the value of $(\partial\rho/\partial c)_\mu$ of the DNA with an RMS error of 0.00527 absorbency units for the 5-scan joint fit, which is shown in Fig. 4. The ratios of the values of A_b as functions of wavelength to the value of A_b at 280 nm for the peptide or A_b at 260 nm for the DNA were then multiplied by the molar extinction coefficient at 280 nm for the peptide or at 260 nm for the DNA to obtain the molar extinction coefficients at the other wavelengths used for taking scans. These values were then used for the construction of the E matrix.

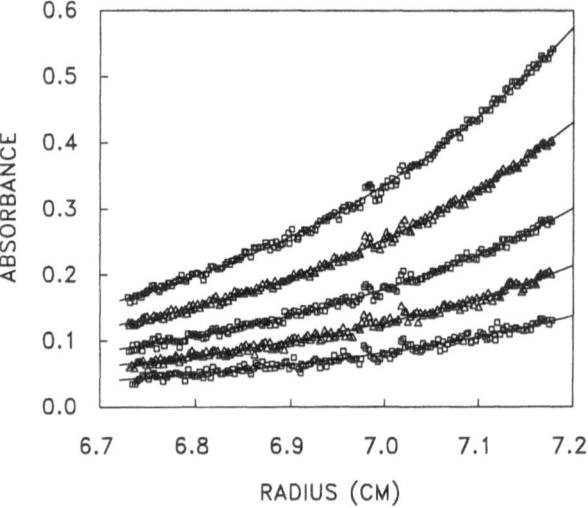

Fig. 3. Distributions of the absorbencies of heat shock peptide at equilibrium at 14000 rev min^{-1} and 4.0°. The scanned wavelengths are, from top to bottom, 230 nm, 232 nm, 234 nm, 236 nm, and 238 nm. The curves are for the joint fit as described in the text. The joint fit RMS error is 0.00520 absorbency units

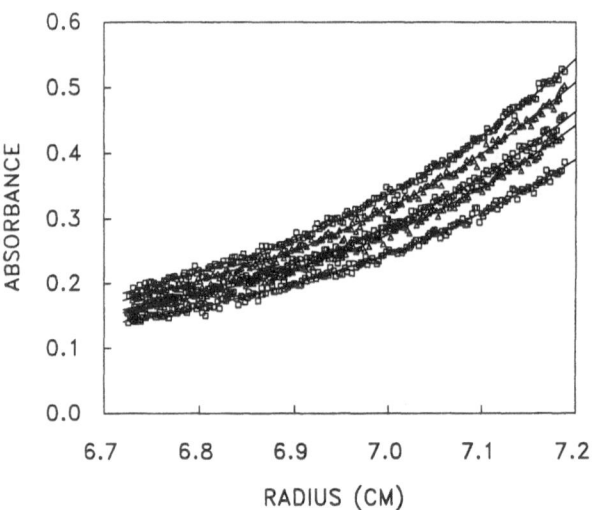

Fig. 4. Distributions of the absorbencies of heat shock DNA at equilibrium at 14000 rev min^{-1} and 4.0°. The scanned wavelengths are, from top to bottom, 246 nm, 244 nm, 242 nm, 240 nm, and 238 nm. The curves are for the joint fit as described in the text. The joint fit RMS error is 0.00527 absorbency units

Since the wavelength selection of the XL-A monochrometer is accurate only to ± 1 nm, it is absolutely essential that all of the measurements to be made at a given wavelength be made without changing the monochrometer setting. Thus, while the wavelength be made without changing the monochrometer setting. Thus, while the wavelength selected may be in error by ± 1 nm, the calculated extinction coefficients will be appropriate for the actual wavelength used for scanning all three solutions. Since the absorption spectra of the peptide and the DNA have relatively broad maxima at 280 nm and 260 nm, respectively, an error of ± 1 nm in actual wavelength will have a minimal effect on the values of the other extinction coefficients. The best results are obtained by using extinction coefficients calculated from data obtained with the XL-A as described above rather than using extinction coefficients calculated from spectrophometric measurements because of this uncertainty of the wavelengths in the XL-A.

Three scans of the equilibrium distribution of the peptide, DNA, and their complex at chemical and centrifugal equilibrium are shown in Fig. 5. The wavelengths are selected to present distributions at the maximum, the minimum, and an average absorbency for the solution. The molar concentration distributions and the fitting lines resulting form a joint fit using Eqs. (8) and (9) as mathematical models are shown in Fig. 6. A value for $\ln K$ of 11.401 ± 0.068 was obtained with an RMS error of 0.0590×10^{-6} M. Figure 7 shows

the distribution of the residuals about the fitting lines. The distributions are quite adequately uniform with no significant systematic deviations, demonstrating not only the quality of the fit, but also that a 1:1 stoichiometry appears to be an appropriate model for the interaction. This value of $\ln K$ agrees well with that obtained in an earlier study where the protein and nucleic acid were

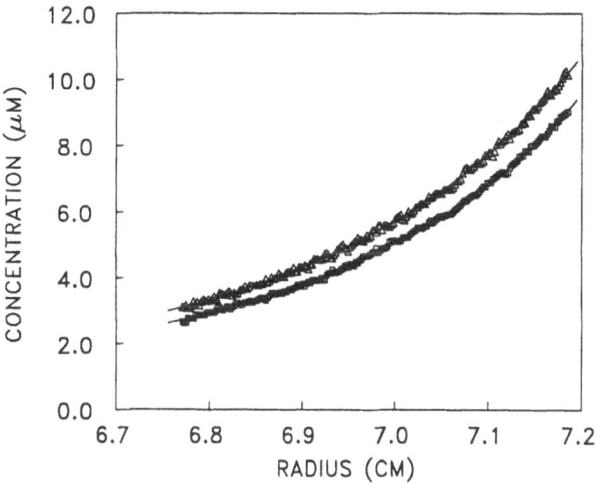

Fig. 6. Molar concentration distributions for the heat shock peptide-DNA interaction. The upper distribution (triangles) is for the peptide plus peptide-DNA complex; the lower distribution (triangles) is for the DNA plus peptide-DNA complex. The curves are for the joint fit as described in the text

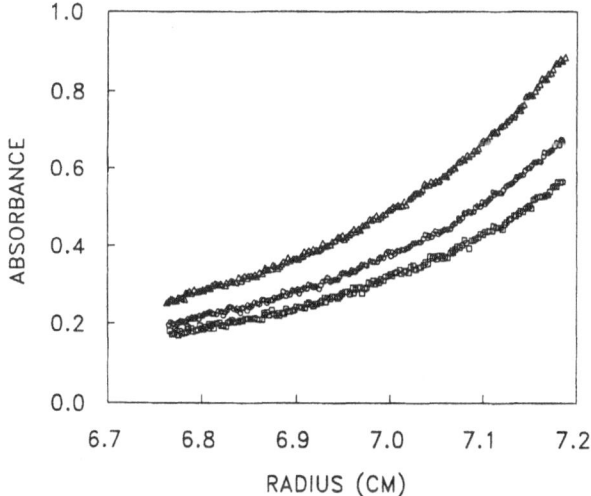

Fig. 5. Distributions of the absorbencies of a mixture of the heat shock factor peptide and DNA in a 2 to 1 molar ratio at equilibrium at 14000 rev min^{-1} and 4.0°. The scanned wavelengths are, from top to bottom, 230 nm, 246 nm, and 238 nm

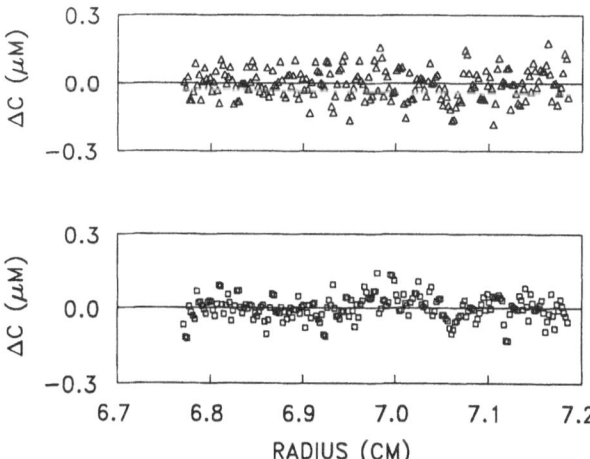

Fig. 7. Distributions of the residuals for the heat shock peptide-DNA interaction shown in Fig. 6. The data symbols are the same as those in that figure. The joint fit RMS error is 0.0590×10^{-6} M

present in a 1:1 molar ratio. Our most recent study using new preparations of both reactants gives a value for ln K of 15.418. Since this value of ln K agrees well with the value of the natural logarithm of the value of K_a obtained by fluorescence titration, it suggests that one or both of the reactants used in the study reported here might be defective.

Measurements of hyperchromism or hypochromism for these reactants have not been made. However, it has been demonstrated that these effects are at a maximum at a wavelength of 265 nm and decrease significantly at shorter wavelengths [18]. A simple calculation show that the value of ln K changes by appproximately ± 0.1 for a $\pm 10\%$ alteration in chromicity, and thus is within the limit of reasonable experimental error.

The validity of the standard error returned by the fitting algorithm of MLAB for this mathematical model and data was investigated by obtaining the sum of squares as a function of ln K by fixing ln K at various values and letting the other parameters attain their optimal values when fitting the data. A graph of the sum of squares as a function of ln K is shown in Fig. 8. With this procedure, if a mathematical model is a linear function of its parameters, and if the data have normally distributed error, and if each datum is properly weighted with the reciprocal of its variance, then the graph of the sum of squares as a function of the parameter of interest will be a parabola that is symmetrical about the optimal value of the parameter. Additionally, the value of the sum of squares at parameter values equal to the optimal parameter value plus and minus one standard error will be equal to the sum of squares at the optimal parameter value plus the square of the root-mean-square error obtained in fitting. This condition was well met in the fitting of these data, and as can be seen in Fig. 8, the condition of symmetry is well met to beyond four standard errors. On the basis of meeting these conditions, we classify the model and data used here as constituting a pseudolinear system. An ongoing Monte Carlo study of such systems indicates that the values of the standard errors obtained during fitting are in reasonably good agreement with the values of the standard errors obtained from the Monte Carlo simulations, generally exceeding them by only a few percent. This is of considerable importance because it validates the use of the

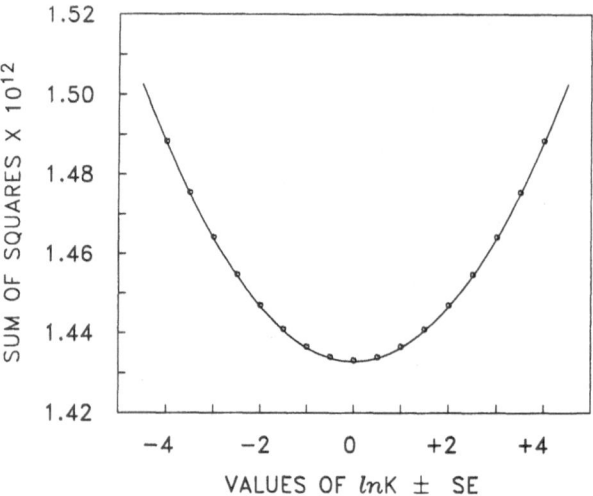

Fig. 8. The sum of squares from fitting the data shown in Fig. 6 as a function of the various fixed values of ln K expressed in terms of increments of standard error for the heat shock peptide-DNA interaction data. The optimal value of ln K is 11.401; the value of the standard error is 0.068

standard errors obtained during fitting for calculating the reciprocals of the variances of the values of ln K to be used as weights when fitting the values of ln K as a function of temperature to obtain values of ΔG^0, ΔH^0, ΔS^0, and ΔC_P^0 for the interactions.

The reliability of this method is primarily attributable to the fact that is utilizes the joint fitting of data matrices of molar concentrations, and these impose very stringent constraints on each other in terms of permitted values of the parameters since both matrices must have identical concentrations of protein-nucleic acid complex as functions of radial position. The use of the pseudoinverse transformation confers additional advantages. The values of the molar concentrations that are obtained by this means are the optimal values obtainable from the absorbency matrix since the multiplication of the E^+ matrix by each row of the A matrix constitutes linear least-squares curve-fitting of that row of the A matrix [16]. Additionally, since the pseudoinverse transformation is linear, error distribution distortion does not occur. These properties have proven useful in the processing of digital images, a situation somewhat analogous to what we are doing here [19].

The application of this method is not limited to protein-nucleic acid interactions. It is equally well suited to the study of protein-protein interactions, provided one of the proteins has a unique chromophore whose absorption spectrum at least partially overlaps the absorption spectra of tyrosine and tryptophan in proteins and also extends beyond them, preferably toward longer wavelengths. In particular, the use of 5-hydroxy tryptophan, which has significantly absorbency to 320 nm and which does not appear to alter biological activity, appears to be particularly promising [20, 21].

Acknowledgement

The authors thank Dr. Carl Wu, Laboratory of Biochemistry, NCI, NIH, for the heat shock DNA binding domain and DNA.

References

1. Riggs A, Suzuki H, Bourgeois S (1970) J Mol Biol 48:67–83.
2. Garner MM, Revzin A (1981) Nucleic Acids Res 9:3047–3059
3. Fried M, Crothers DM, (1981) Nucleic Acids Res 9:6505–6525
4. Strauss F, Varshavsky, A. (1984) Cell 37:889–901
5. Wissman A, Hillen W (1991) In: Sauer, RT (ed) Methods in Enzymology, Vol 208: Protein-DNA Interaction. Academic Press, New York, pp 365–379
6. Galas DG, Schmitz A (1978) Nucleic Acids Res 5:3157–3170
7. Dixon, WJ, Hayes JJ, Levin JR, Weidner MF, Dombroski BA, Tullius TD (1991) In: Sauer, RT (ed) Methods in Enzymology, Vol. 208: Protein-DNA Interaction. Academic Press, New York, pp 380–413
8. Dervan P (1991) In: Sauer RT (ed) Methods in Enzymology, Vol. 208: Protein-DNA Interaction. Academic Press, New York, pp 497–515
9. Koblan KS, Bain DL, Beckett D, Shea MA, Ackers G (1992) In: Brand L, Johnson M (eds) Methods in Enzymology, Vol. 210: Numerical Computer Methods. Academic Press, New York, pp 497–515
10. Draper DE, Von Hippel PH (1978) J Mol Biol 122:321–338
11. Lundback T, Cairns C, Gustafsson JA, Carlstedt-Duke J, Hard T (1993) Biochemistry 32:5074–5082
12. Shamoo Y, Ghosaini LR, Keating KM, Williams KR, Sturtevant JM and Konigsberg WH (1989) Biochemistry 28:7409–7417
13. Lewis MS, Kim S-J, Kumar A, Wilson SH (1992) Biophys J 61:A489
14. Schmidt B, Riesner D (1992) In: Harding SE, Rowe AJ, Horton JC (eds) Analytical Ultracentrifugation in Biochemistry and Polymer Science. Royal Society of Chemistry, Cambridge, pp 176–207
15. Lewis MS (1991) Biochemistry 30:11716–11719
16. Strang G (1986) In: Introduction to Applied Mathematics. Wellesley-Cambridge Press, Wellesley, MA, pp 138–139
17. Durchschlag H (1986) In: Hinz H-J (ed) Thermodynamic Data for Biochemistry and Biotechnology. Springer-Verlag, New York, pp 46–50
20. Jansen DE, Kelly RC, von Hippel PH (1976) J Biol Chem 251:7215–7228
19. Pratt WK (1978) In: Digital Image Processing. John Wiley and Sons, New York, pp 206–211
20. Ross JBA, Senear DF, Waxman E, Kombo BB, Rusinova E, Huang YT, Laws WR, Hasselbacher CA (1992) Proc Nat Acad Sci USA 89:12023–12027
21. Hogue CWV, Rasquinha I, Szabo A, MacManus JP, (1992) FEBS 310:269–272

Received June 19, 1993;
Accepted September 23, 1993

Authors' address:

Marc S. Lewis
Biomedical Engineering and Instrumentation Program
National Center for Research Resources
National Institutes of Health
Bethesda, MD 20892 USA

Progress in Colloid & Polymer Science

Progr Colloid Polym Sci 94: 54–65 (1994)

Sedimentation equilibrium analysis of glycopolymers

S. E. Harding

University of Nottingham, School of Agriculture, Sutton Bonington, United Kingdom

Abstract: The glycopolymers – a general term used to represent polysaccharides and glycoconjugates collectively – present the analytical ultracentrifuge – and in particular sedimentation equilibrium analysis – with one of its greatest challenges. In this paper the difficult nature of these substances will be described as well as why the inherent fractionation nature of the sedimentation equilibrium method gives it an edge over other techniques. The problems of limited choice of optical system which can be applied (through lack of naturally occurring chromophores), the importance of both the Rayleigh and Schlieren optical systems for these substances, the inapplicability of the meniscus depletion method, how we can get meniscii concentrations out, automatic data capture and analysis, extraction of "whole distribution" and point average molar masses, coping with the severe non-ideality one often finds with solutions of these substances and how we can distinguish between "polydispersity" (i.e., the presence of species of different molar mass or density not in chemical equilibrium) and self-association are all discussed. Finally, and after taking onboard the non-ideality problem, four methods of extracting distributions of molar mass are considered, focussing on one method which combines sedimentation equilibrium with gel permeation chromatography. In concluding, this theme of the importance of combining data with that from other techniques is continued by discussing the important relation sedimentation equilibrium has with classical (i.e., so-called "static") light scattering procedures.

Key words: Polysaccharides – mucins – molar mass averages – molar mass distributions – Rayleigh optics – Schlieren optics – relation of other techniques

Introduction

Inasmuch as analytical ultracentrifugation – particularly sedimentation equilibrium – has become a Cinderella technique amongst biochemical and polymers scientists – compared to more fashionable techniques like gel permeation chromatography, gel electrophoresis, mass spectrometry, NMR, and light scattering – glycopolymers or "polyhydroxycompounds" (polysaccharides and glycoconjugates) are very much the Cinderella molecules of biochemistry when compared to the more fashionable proteins and nucleic acids. This paper is therefore about characterising Cinderella molecules by a Cinderella technique.

It has been interesting to observe both molecule and technique now emerging from their Cinderella status, largely through a growing appreciation of the importance of glycopolymers in molecular recognition phenomena on the one part, and a growing appreciation of the limitations of other techniques – techniques which lack the absolute and inherent fractionation nature of sedimentation methods on the other. This is particularly relevant to the characterisation of difficult heterogeneous macromolecules of which glycopolymers are only one class but which possibly present the ultracentrifuge with one of its biggest challenges (see, e.g., Ref.[1]). In this paper I am going to describe some of the many problems we have encountered in trying to use sedimentation equilibrium in the analytical ultracentrifuge to study the size, size distribution and interactions of these molecules focussing on polysaccharides and a particularly difficult class of polysaccharide-protein conjugate known as either the "muco-polysaccharides", "mucus glycoproteins" or more simply "mucins"[2].

Although the vast majority of analytical ultra-centrifuge users have no direct interest in these molecules in particular, others might possibly pick up a few tips or clues as to how to apply sedimentation equilibrium procedures to difficult heterogeneous macromolecules in general of which the glycopolymers are only one small – but nontheless interesting – class.

The challenge

A typical glycopolymer can have a poorly de-fined conformation in solution with a large ca-pacity to trap and entrain surrounding solvent molecules with a resulting *high exclusion volume* (sometimes > 100x in excess of the anhydrous volume) and hence high thermodynamic non-ideality. Another contribution to thermodynamic non-ideality can arise from *polyelectrolyte behavi-our* if the molecule has a high unsuppressed charge in solution, such as a pectin in a low ionic strength solvent: many "food grade" polysacchar-ides for example are polyannionic. Mucins are polyanionic because of their sialic acid content. Glycopolymer preparations are usually *polydis-perse*, that is, they contain species of different molar mass (either in a "discrete"-viz. paucidis-perse-or quasi-continuous sense). Finally some can have the ability to *self-associate* in solution, such as guar. The net result is that these molecules place a considerable strain on the available ul-tracentrifuge methodology, and in many cases satisfactory information can only be achieved by combining results with other techniques, notably light scattering, gel permeation chromatography or electron microscopy (see, e.g., Ref.[3]).

The basic information sought and the problems

Leaving aside sedimentation velocity and ana-lytical density gradient analysis, what is the sort of basic information we are after using sedimenta-tion equilibrium analysis on these molecules? And of course, it is in terms of molar mass analysis in the form of either molar mass averages (both "point" and "whole cell"), molar mass distribu-tions and, if the system performs in complexation phenomena or self-association reactions, in terms of stoichiometries. But there are a number of problems we have to overcome or "troubleshoot", such as:

1) a limited choice of optical system;
2) the inapplicability of the "meniscus de-pletion method";
3) the evaluation of the meniscus concentration from Rayleigh optical records;
4) strong curvature in the log concentration versus distance squared plots;
5) strong thermodynamic non-ideality and misleading pseudo-ideality;
6) distinguishing polydispersity (i.e. species of different molar mass not in chemical equili-brium) from self-association.

Limited choice of optical system

Most polysaccharides and many glycoconju-gates do not contain sufficient chromophore in the visible or near (i.e. "useable") part of the ultra-violet so, in general, we cannot use the absorption optical system but instead either of two refrac-tometric methods: Schlieren optics or Rayleigh interference*. The latter is preferable because of its greater sensitivity at lower concentration, but the optical record is one of solute concentration *relative* to the meniscus, rather than absolute con-centration directly. For simple experiments on proteins a popular way of avoiding this problem is to use the meniscus depletion method [5], where the ultracentrifuge is run at a high enough speed so the meniscus is effectively depleted of

---

*) The new XLA ultracentrifuge from Beckman instruments [4] appears to have stable optics in the far ultraviolet (200–230 nm). All mucins and many polysaccharides absorb significantly in this region and we are exploring the use of "far-uv" detection, with appropriate baselines, for both sedimentation equilibrium and sedimentation velocity.

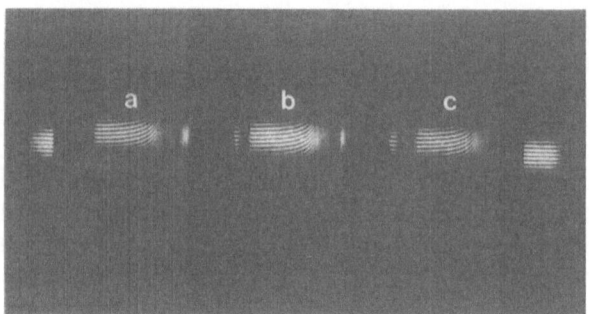

Fig. 1. Rayleigh equilibrium interference patterns for a mucin (bronchial mucin BM GRE) in three different solvents a) a phosphate/chloride buffer containing 0.4 M CsCl; b) a phosphate/chloride buffer containing 5 mg/ml fucose; c) a phosphate/chloride buffer containing 5 mg/ml N-acetyl-glucosamine. The initial mucin cell loading concentration in each case was ~ 0.4 mg/ml (30 mm path length cell). The rotor speed was 1967 rev/min. From Ref. [6]

macromolecular solute – the optical record is then one of absolute concentration – in fringe number or weight terms – versus radius.

Inapplicability of the "high speed" or "meniscus depletion" method

Figure 1 illustrates the next problem any sedimentation equilibrium analysis on these materials needs to overcome: that is we cannot use the "meniscus depletion" (see, e.g., Ref [5]), method, widely used in protein biochemistry. Figure 1 in fact shows the solution Rayleigh fringes from a low-speed sedimentation equilibrium experiment 'on a mucin of mass average molecular mass ~ 6 million, and at low loading concentration (~ 0.4 mg/ml), run in three separate solvents. Because of the polydispersity of these materials it is generally impossible (except in cases of pseudo-non-ideality) to choose run conditions to get proper meniscus depletion without losing optical registration of the fringes near the cell base: one can observe clearly in Fig. 1 the steep rising fringes at the cell base but finite slope of the fringes near the meniscus.
There is a further problem with the high speed meniscus depletion method in terms of a speed dependent enhancement of the effective thermodynamic second virial coefficient, B_{eff}, as the

following equation shows [7]

$$B_{eff} = B \left\{ 1 + \frac{\lambda^2 M_z^2}{12} + \cdots \right\}, \qquad (1)$$

where λ is a function of the *square* of the rotor speed ($\lambda = (1 - \bar{v}\rho)\omega^2(b^2 - a^2)/2RT$, $\bar{v}$ being the partial specific volume, ρ the solution density, ω the angular velocity and a and b the radial positions at the cell meniscus and base respectively). The collective result is that with polysaccharides we have to use the low or intermediate speed method with the requirement of a method for evaluating the concentration at the meniscus, either in terms of g/ml or in terms of fringe numbers. The next problem is thus: how do we get meniscii concentrations out?

Evaluation of the meniscus concentration

My old mentor J.M. Creeth produced in the late 1960s with R. Pain [8] a very useful review of all the methods for getting meniscii concentrations (denoted "C_a" in weight concentration terms or, more usually "J_a" in the equivalent fringe number terms where the subscript a represents the radial position at the meniscus) out; however, we found the most useful method was one where we get J_a by some simple graphical manipulation of the basic fringe data [9]. We can get J_a usually to an accuracy of a few percent from the ratio of twice the intercept to the limiting slope of a plot of $j/(r^2 - a^2)$ against $\{1/(r^2 - a^2)\}\int_a^r rj dr$ (Fig. 2), where j is the concentration in fringe numbers ("fringe concentration") *relative to the meniscus* and r is the radial displacement. The method usually gives J_a to an accuracy of the order of 0.1 fringe. We find an adapted sliding strip procedure useful for this purpose [10], especially where the plots are strongly curved as in Fig. 2b. Strong curvature is a symptom of either non-ideality or heterogeneity (the sense of the curve depends on which is the stronger effect).

Automatic "off-line" data capture

For getting out J_a using this method multiple data collection and averaging is very important for strongly curving systems such as shown in

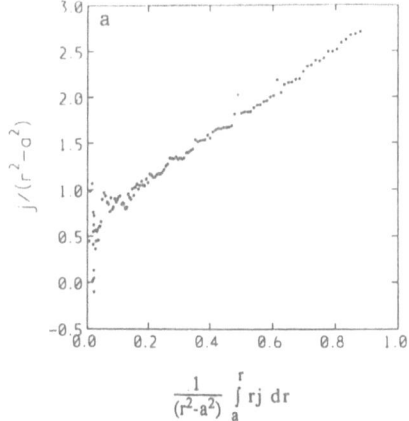
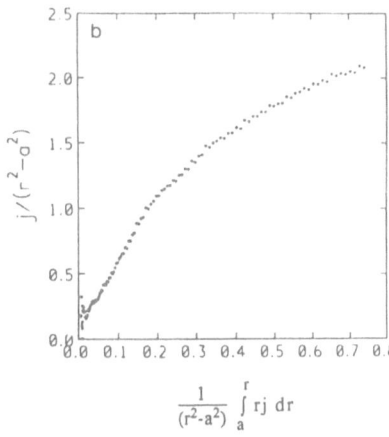

Fig. 2. Extraction of meniscus concentration using the method of ref. [9] for a) a fairly homogeneous/ideal solution of colonic mucin "T-domains" $\{J_a \sim 0.58 \pm 0.05\}$; b) a highly non-ideal solution of xanthan ("RD") ($J_a \sim 0.01 \pm 0.01$, i.e., near depletion conditions). From Ref. [10]

Fig. 3. LKB (Bromma, Sweden) Laser Densitometer set up at Nottingham. This is used to capture (into the Amstrad PC) automatically off-line from photographic film our Rayleigh interference data. The Fourier cosine series algorithm "ANALYSER" produces a 100–200 pt concentration versus distance dataset (as shown in the inset for a low speed sedimentation equilibrium experiment on a commercial guar sample) which is transferred to the FORTRAN programme MSTAR (10) on the mainframe IBM 3081/Q for full molecular weight analysis

Fig. 2b, and for this purpose the use of automatic multiple data capture and analysis is of extreme value here. We capture our data automatically but not directly *on-line*, as described by, for example, Laue [11] but *off-line*. That is to say we take a photograph and digitise it using a laser densitometer of the sort you can find in many biochemistry departments (Fig. 3) – these things are normally used for scanning SDS gels, and we use a simple Fourier cosine series algorithm to average over the fringe data set to give our sedimentation equilibrium concentration distribution, to an accuracy comparable with the on-line set-up that Laue has described (see, e.g., Ref. [12]).

Extraction of average molar masses

Whole cell mass average

The next problem is in getting molar masses out – even getting whole cell or whole distribution mass average values can be much more tricky compared with getting the same for simple well behaved protein systems. Plots of the logarithm of the concentration against radial displacement squared are often strongly curved because of either heterogeneity effects (Fig. 4a) or nonideality effects (unless by a lucky coincidence the results of the two effects cancel to give a

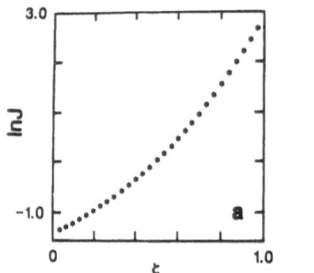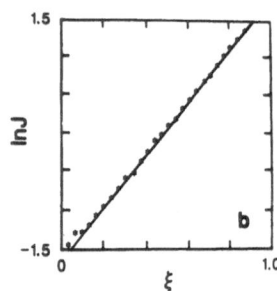

Fig. 4. Log concentration versus distance squared data evaluated from Rayleigh equilibrium optical records for two mucins a) bronchial mucin (from a cystic fibrosis patient) BM GRE, b) bronchial mucin (from a cystic fibrosis patient) CF PHI. ξ is a normalised radial displacement squared parameter, normalised so it has a value of 0 at the meniscus and 1 at the cell base: $\xi = \{r^2 - a^2\}/\{b^2 - a^2\}$. Note the strong curvature of a) and the pseudo-ideality of b). (see, e.g., Ref. [10])

"pseudo-ideal" profile, as in Fig. 4b). Whatever, to get the whole distribution mass average you need to estimate the concentration or the logarithm of the concentration not only at the meniscus, but also *at the cell base*, and this can be very tricky, especially in the case of strong curvature (e.g., Fig. 4a) or if the base is not well defined. One way of minimising this problem is to use a function known as "M^*" defined by [9]

$$M^*(r) = \frac{j}{kJ_a(r^2 - a^2) + 2k\int_a^r rj dr}, \qquad (2)$$

where a and b are the radial positions at the solution meniscus and cell base respectively and k is the usual constant [1] in terms of rotor speed, ω, solvent density, ρ_0[†], and partial specific volume $\bar{v}$:

$$k = \frac{(1 - \bar{v}\rho_0)\omega^2}{2RT}. \qquad (3)$$

J.M. Creeth and myself [9] found M^* defined in such a way to have some useful properties: the most important of these is that its value extrapolated to the cell base equals the weight average over the whole solute distribution "$M^0_{w,app}$" (where the "0" signifies its over the whole cell and the "app" signifies an apparent value at a finite cell loading concentration) and, as we found by extensive simulations [1, 9], for five different types

of systems the M^* method appeared to represent a considerable improvement for extracting $M^0_{w,app}$ compared to the conventional "average slope" or log concentration extrapolation methods [8].

Point mass average molar masses ("$M_{w,app}$")

These are relatively straightforward to produce from the fringe concentration data as $d\ln J/dr^2$ times $(1/k)$, (where $J = j + J_a$ and k is as defined in Eq. (3)), but again this depends on a reasonable estimate for J_a. ΔJ (i.e., the difference in fringe concentration between the meniscus and cell base) needs to be at least four fringes for the $M_{w,app}$ data to be reliable without heavy smoothing. We use sliding strip procedures along the lines discussed by Teller [13], with an 11pt sliding strip for a total data set of 100–200 radial positions.

Number and z- whole cell averages

The number point average molar mass, $M_{n,app}$ you can in my opinion forget about for these substances (unless, for pseudo-ideal systems, you can use the meniscus depletion method): besides J_a you also need to estimate $M_n(a)$ [13]: the same problem applies to the whole-cell number average. The situation is not quite so bad for z-averages. If Rayleigh optics are used the point z-average is independent of errors in J_a, although it depends on the ratio of a double differential to a single differential viz, data of very high precision is necessary; the whole cell z-average requires accurate estimates of not only J_a and J_b but also the point mass averages at the meniscus and base (see e.g., Refs, [13, 10]). If reliable z-averages are required then Schlieren optics should be used, which yield $M^0_{z,app}$ and $M_{z,app}$ directly via the Lamm equation (see, e.g., Ref. [8]), and at concentrations now claimed as low (using the Fresnel fringes, Ref. [12]) as can be achieved from Rayleigh fringes: this would therefore be my method of choice. Indeed, at Nottingham, although we have two Model E's with laser light sources dedicated to Rayleigh optics, we have another dedicated to the Schlieren system producing M_z information.

[†]) Nowadays the solvent density appears preferable over the solution density in this term (see, e.g., Ref. [1] & references cited therein).

The problem of thermodynamic non-ideality

If I can now focus on the problem of the non-ideality of these substances, the first feature to be aware of is that the symptoms of heterogeneity (which tend to produce upward curvature in the log concentration versus distance squared plots), as in Fig. 4a, can sometimes apparently cancel the symptoms of non-ideality which produce downward curvature to give a pseudo-ideal monodisperse linear plot, as shown in Fig. 4b for a mucin, which can be very misleading. The "pseudo-ideal" system corresponding to Fig. 4b is in fact both very polydisperse and very non-ideal, and this feature illustrates a statement made long ago by Teller [13] that a linear plot of $\log c$. versus distance squared from a single sedimentation equilibrium experiment is not by itself sufficient evidence for monodispersity or ideality.

And, as the Gilberts [15] and others have shown, since a single symmetric boundary from a sedimentation velocity experiment is also insufficient criterion for solute homogeneity, to this end Creeth and myself developed a simple assay for solute homogeneity using two cells of different optical path length in a multihole rotor [16]. The idea is to compare solution interference fringes of the same initial loading concentration if expressed on a *fringe number* basis but different loading concentrations expressed on a *weight* basis. This test can be performed in a single sedimentation equilibrium experiment if a multi-hole rotor is used together with a suitable combination of two cells (double sector, one with wedge window if the instrument does not have a multiplexing system, or multichannel). One cell or pair of interference channels has say a 30 mm optical path length, the other only a 12 mm path length but 2.5 times higher mass concentration to compensate: fringe patterns, and corresponding average molar masses will be identical *only* for a homogeneous ideal system, and not otherwise. In this way the mucopolysaccharide CFPHI, which *appeared* homogeneous and ideal from the linear log concentration versus distance squared plot of Fig. 4b can clearly be shown to be otherwise (Fig. 5). These observations clearly illustrate the dangers of inferring solute homogeneity or ideality from a single log concentration versus distance squared plot. Some (non-polysaccharide) systems do "pass the test" however, such as a dilute solution of the small virus TYMV [17].

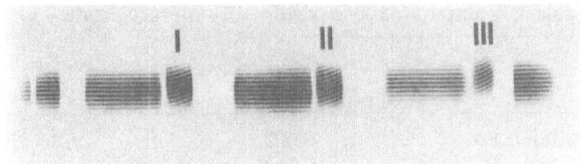

Fig. 5. Sedimentation equilibrium sample homogeneity test. Comparison of Rayleigh equilibrium patterns of the same initial loading concentration on a fringe number basis but differing initial loading concentration on a mass (mg/ml) basis, c^o. Patterns I and II: $c^o = 3.0$ mg/ml, 12 mm path length multi-channel cell (inner two pairs of channels used, outer pair masked) pattern III: $c^o = 1.2$ mg/ml, 30 mm path-length multi-channel cell (outer pair of channels used, the inner two pairs masked off). Note the different curvature of III confirming the system is neither homogeneous nor ideal. The identical nature of I and II confirms that the differing radial positions of each does not affect their concentration distribution. The solution column of III appears shorter because i) the apparent width of the meniscus is much greater in the cell of longer optical path length and ii) the fringes at the cell base are steeper and accordingly less intense and so are partially lost on photographic reduction (from Ref. [16]

Moving back to the non-ideality problem this can be very severe for polysaccharides as Table 1 shows. The interesting column is the "$1 + 2BMc$" one and represents the factor by which an apparent molecular weight, measured at a concentration as low as 0.2 mg/ml, underestimates the true or infinite dilution value. For many, of course, the effect is not too bad – less than a few percent, but for some like alginates, serious error can result – an underestimate of over 40% for example. And for cases like these it is necessary to measure the apparent molar mass – whether it be mass or z-average, at a number of concentrations and extrapolate to zero in the standard way (see, e.g., Ref. [24]). In extreme cases – such as some alginate preparations, even under conditions of ionic strength where polyelectrolyte behaviour should be largely suppressed, the two virial coefficients ($1/M$ and B) are not sufficient to account for the concentration behaviour, even under dilute solution conditions [28].

Non-ideality also reveals itself in *point average* representations of the data as shown in Fig. 6 for dilute solutions of a neutral but tricky polysaccharide know as guar gum. For many polysaccharides like this we observe a maximum in the point average versus fringe concentration data, both in terms of mass average (Fig. 6a) and in terms of the point z-average (Fig. 6b) and the

Table 1. Comparative non-ideality of polysaccharides

	$10^{-6} \times M$ g/mol	$10^{-4} \times B$ ml·mol/g²	BM ml/g	[a])$1 + 2BMc$	Ref.
Pullulan P5	0.0053	10.3	5.5	1.002	18
Pullulan P50	0.047	5.5	25.9	1.010	18
Xanthan (fraction)	0.36	2.4	86	1.035	19
β-glucan	0.17	6.1	104	1.042	20
Chitosan (KN-50-1)	0.064	1.7	109	1.044	21
Dextran T500	0.42	3.4	143	1.057	22
Pullulan P800	0.76	2.3	175	1.070	18
Chitosan (Protan 203)	0.44	5.1	224	1.090	23
Pullulan P1200	1.24	2.2	273	1.109	18
Bronchial mucin CFPHI	2.0	1.5	300	1.120	24
Chitosan (Protan SeaCure)	0.16	27.5	445.5	1.178	21
Pectin (citrus fraction)	0.045	50.0	450	1.180	25
Scleroglucan	5.7	0.50	570	1.228	26
Alginate	0.35	29.0	1015	1.406	27

[a]) at a loading concentration of 0.2 mg/ml

existence of these maxima is symptomatic of a heterogeneous but highly non-ideal system, with the effects of polydispersity dominating at low radial positions and non-ideality effects dominating at the higher positions.

The problem of distinguishing polydispersity from self-association

Figure 7 [30] shows another point mass average versus concentration plot for a muco-polysaccharide, again with a maximum, again characteristic of a heterogeneous but highly non-ideal system. The question of interest is what is the prime source of the heterogeneity? Is it because of the presence of components of different molecular weight that are not-interacting (i.e., not in chemical equilibrium), a phenomenon we call "polydispersity", or is self-association behaviour the main contributor?: many polysaccharides, such as guar and mucopolysaccharides are thought to self-associate, and there is now increasing interest in the food and pharmaceutical fields of possible inter

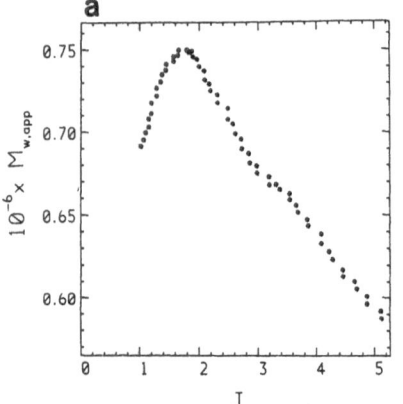

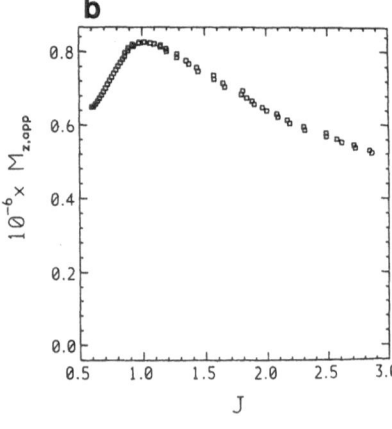

Fig. 6. Plots of (apparent) point M_w, M_z versus fringe number concentration J from a low speed sedimentation equilibrium experiment on a purified guar preparation. Rotor speed = 5200 rev/min, loading concentration $c^o \sim 0.7$ mg/ml. Rayleigh interference optics. From Ref. [29]

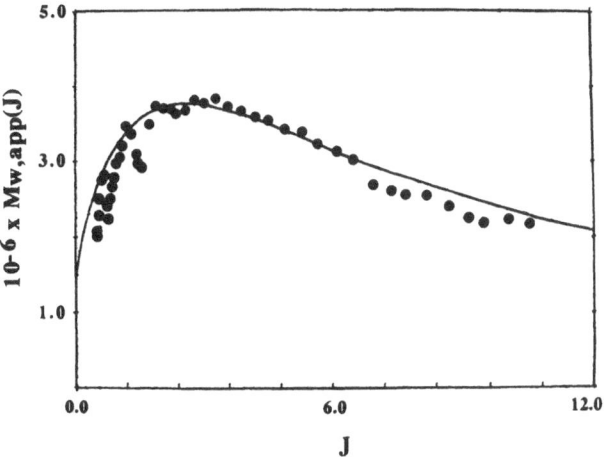

Fig. 7. Plot of (apparent) point M_w versus fringe concentration (with respect to a 12 mm optical path length cell) from a low speed sedimentation equilibrium experiment on bronchial mucin BM GRE (loading concentration, $c^o = 1.0$ mg/ml, solution column length = 3 mm) The line fitted corresponds to an effective non-ideal isodesmic self-association with monomer molar mass $M_1 = 1.5 \times 10^6$ g/mol, second virial coefficient, $B = 0.33 \times 10^{-6}$ mol.ml^{-1}fringe^{-1}, $k = 1.2$ fringe^{-1} (see, e.g., Ref. [30] where similar fits are given but with respect to a 30 mm path length cell)

actions in mixed polysaccharide systems [31]. Both effects produce upward curvature (i.e., positive second differential) in plots of log concentration versus distance squared or point average molar mass versus concentration. Initially when we worked on mucopolysaccharides or mucins we thought that the prime cause was the latter [32] and indeed we get an excellent fit to the observed point average data if as a first approximation we

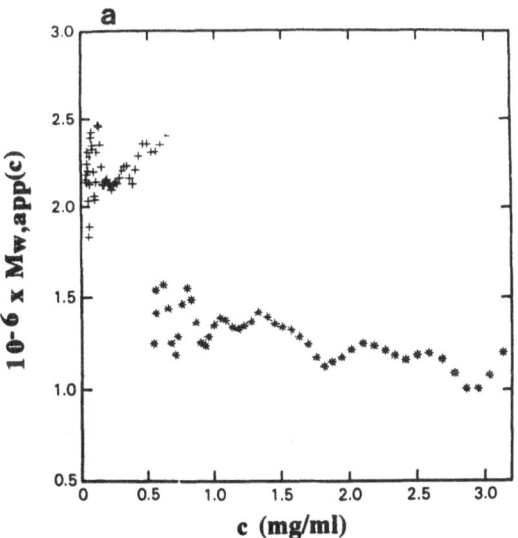

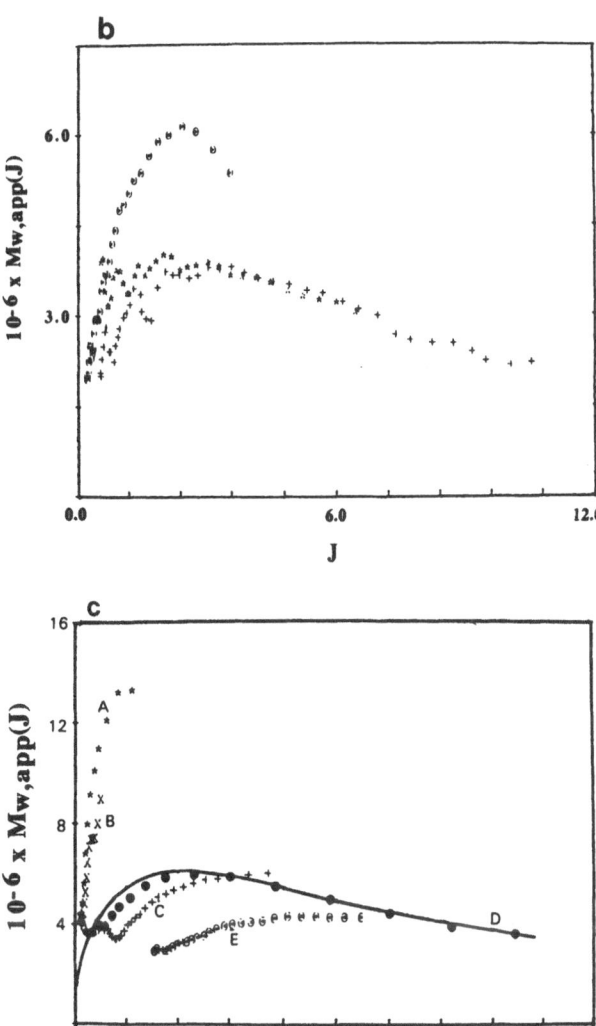

Fig. 8. Sedimentation Equilibrium Polydispersity assay. Different loading concentration "non-overlap" plots of point mass average apparent M_w versus concentration for three mucins a) Bronchial mucin CF PHI $+$: c^o(initial loading concentration in mg/ml) ~ 0.2 mg/ml, 30 mm cell; *: $c \sim 2.0$ mg/ml, 12 mm cell Both with 3 mm solution columns. From Ref. [6]. b) Bronchial mucin BMGRE $\odot$: c^o 0.4 mg/ml, 30 mm cell (fringe concentrations corrected to the equivalent values in a 12 mm cell) *: $c^o \sim 0.7$ mg/ml, 12 mm cell, $+$: $c^o \sim 1.0$ mg/ml, 12 mm cell. All three data sets correspond to 3 mm solution columns. From Ref. [30]. c) Pig gastric mucin *: J^o (initial loading concentration in fringe numbers) ~ 0.32, column length 3 mm; x:$J^o \sim 0.42$, 1.5 mm; $+$: $J^o \sim 3.19$, 1.5 mm; $\bullet$: $J^o \sim 4.03$, 3 mm;$\odot$: $J^o \sim 6.00$, 1.5 mm. All three data sets correspond to a 30 mm optical path length cell. The line fitted to data set in (c) corresponds to an effective non-ideal isodesmic association (see legend to Fig. 8), with $M_1 = 1.5 \times 10^{-6}$, $B = 0.013 \times 10^{-6}$ mol.ml^{-1}fringe^{-1} and $k = 2.1$ fringe^{-1} where the fringe concentration units this time refer to a 30 mm path length cell. From Ref. [30]

ignore polydispersity and assume a self-association: Ref. [30] gives one example and another is given in Fig. 7 for the bronchial mucopolysaccharide "BM GRE": the fit given in this figure corresponds to a non-ideal isodesmic (i.e., each monomer is added on with constant free energy increment) self-association with plausible values for the isodesmic association constant, k, the "monomer" molar mass, M_1 and the second virial coefficient B. We can also model the log concentration versus distance squared data directly (the line fitted in Fig. 4b for the mucin "CFPHI" corresponds to an indefinite isodesmic association with $M_1 = 2.15 \times 10^6$ g/mol, $k = 260$ ml/g and $B = 1.5 \times 10^{-4}$ ml.mol.g^{-2}). Despite the good fits we know from other measurements that both BM GRE and CF PHI *are not interacting at all*, illustrating another pitfall we can easily fall into, viz. the effects of polydispersity of these types of system cannot be ignored. There are diagnostic procedures available to assay whether polydispersity effects *can* be ignored: Roark & Yphantis have shown [33] that for a purely non-ideal self-associating system plots of point mass average molecular mass versus concentration for differing cell loading concentrations should superimpose. So for a simple self-association such as lysozyme they do [34], but for mucins for example they do not as the three examples in Fig. 8 show. In fact further experiments comparing fringe profiles and corresponding molar masses for these molecules in dissociative and non-dissociative solvents (in the case of mucins by adding 6 M GuHCl or swamping concentrations of fucose, galactose or N-acetyl glucosamine) and observing the lack of any effect of the latter (Fig. 1) [2, 6] have shown that for mucopolysaccharides in dilute solution self-association phenomena is negligible – the observed heterogeneity is due virtually entirely to heterogeneity of components not in chemical equilibria-i.e., polydispersity.

Extraction of molar mass distributions

The profiles in Figs. 6–8 are not molar mass distributions and in trying to get this sort of information out once again the main stumbling block we have to overcome is that of thermodynamic non-ideality. There are four possible routes open to us here (Table 2).

Table 2. Molar mass distribution analysis by low speed sedimentation equilibrium

Method	Type of analysis	Ref
I	Polydispersity indices (M_z/M_w etc)	35
II	Non-ideal-polydisperse modelling of log concentration versus distance squared data	13,34
III	Equivalent self-association fit	28
IV	Off-line coupling to gel permeation chromatography	37

Method I: Polydispersity indices

The simplest way is by using the ratios of whole cell averages, or polydispersity indices, which is fine so long as you can measure numbers or z-averages to a reasonable precision *after* correction for non-ideality. Number averages can only be extracted with ease from Rayleigh patterns if the high speed method is used [5] – thereby effectively ruling them out for polysaccharides for the reasons given above. z-averages can be obtained with relatively high precision using Schlieren optics and now at concentrations as low as can be obtained using Rayleigh optics [12] (or from Rayleigh records themselves but at much lower precision-see above). These ratios can either be used directly as so-called "polydispersity indices" or related to the standard deviation of a distribution (whatever form this may take) via special relations known as the "Herdan relations" ([35], see also refs. [2, 8]).

Method II. Non-ideal-polydisperse modelling of log concentration versus distance squared data

The more direct way is to model directly the log concentration versus radial displacement plots by fitting the parameters of a non-ideal polydisperse system. Although this method came out in the Biophysical Journal over 8 years ago now [14], because of the particularly complex interdependent nature of the non-linear equations involved *for the low speed case* – largely caused by the non-ideality term-it takes a great toll on computer resources, even on the fastest computers around such as the IBM 3081 at Cambridge. So, we have presently been unable to apply it to quasi-continuous distributions of molar mass that are, for example, the hallmark of polysaccharides, but

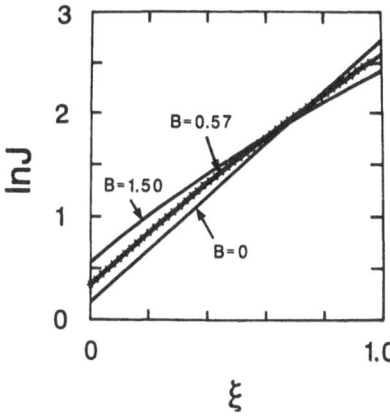

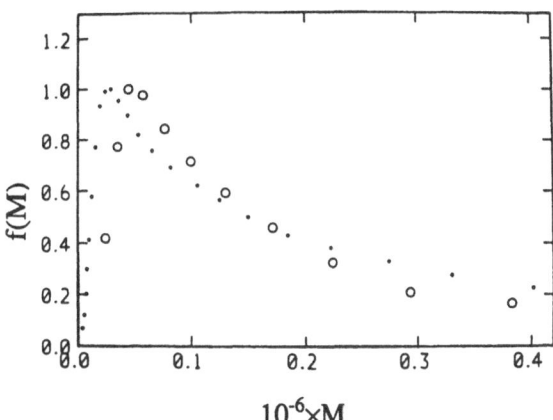

Fig. 9. Molar mass distribution modelling of bronchial mucin CF PHI by Method II. Non-ideal three-component fit to the observed log concentration versus distance squared data for a low speed sedimentation equilibrium experiment. The line fitted corresponds to the following parameters for each of the three components: component 1, $M_1 = 1.2 \times 10^6$ g/mol, $J^\circ_1 = 0.9$ fringe; component 2, $M_2 = 1.8 \times 10^6$ g/mol, $J^\circ_2 = 3.6$ fringe; component 3, $M_3 = 2.4 \times 10^6$ g/mol, $J^\circ_3 = 0.9$ fringe. B has been expressed as its value $\times 10^4$ (ml·mol·g^{-2}). The three component model is based on observations from platinum-shadowed electron microscopy. From Ref. [14].

Fig. 11. Molar mass distribution of citrus pectin, evaluated according to Method IV (open circles). The filled circles corresponds to the distribution on the same material evaluated by classical light scattering procedures coupled to gel permeation chromatography. Method IV is our method of choice. From Ref. [3]

tion distribution for the non-ideal case have been further examined by Lechner [36].

Method III. Equivalent self-association fit

A much easier way although theoretically less elegant than the previous method, is to use to our advantage the property of indistinguishability *from a single experiment* between a non-ideal polydisperse system and a non-ideal self-associating system. It is therefore possible to apply the much easier to handle equations of, for example, a non-ideal isodesmic association to calculate a constant which, when applied to a static system will define a distribution of molar mass, no matter what the cause of the distribution is [30, 2] and again, this has been successfully applied to mucins (Fig. 10).

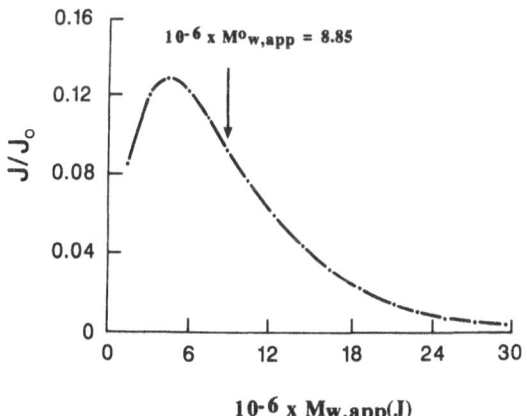

Fig. 10. Molar mass distribution of pig gastric mucin, evaluated according to Method III. The distribution corresponds to the fit shown in Fig. 8c. The value marked by an arrow corresponds to the weight average (apparent) molecular weight for the whole solute distribution. From Ref. [30] (see also Ref. [2])

Method IV. Off-line coupling to gel permeation chromatography

From a practical point of view, we find the best procedure is to use sedimentation equilibrium in conjunction with gel permeation chromatography (gpc) to provide an absolute calibration for the latter [37]. The idea is to isolate fractions of narrow bandwidth (in terms of elution volume) from the gpc eluate, determine their molecular weights using sedimentation equilibrium using short solution columns (0.7 mm–1.5 mm) and

nontheless successfully applied to discrete distributions of molar mass, which at least partially represent mucins [2, 14] (Fig. 9). The problems associated with the modelling of the concentra-

multi-channel cells to speed things up, thereby giving an absolute calibration for the gpc columns. Distributions of molecular weight found in this way have been in remarkable agreement with similar procedures involving light scattering as shown in Fig. 11: Method IV is in my opinion the method of choice. It is called "off-line" because the eluate is not fed directly into the ultracentrifuge cell whilst the ultracentrifuge is running (this distinguishes itself from certain light scattering photometers which are "on-line"-i.e., directly connected between gpc columns and a concentration-usually refractive index based-detector).

Concluding remarks: the relation of sedimentation equilibrium with "static" light scattering

We have just seen two examples of the importance of combining sedimentation with other data – in the cases above it was with electron microscopy and gel permeation chromatography. This importance cannot be overstressed for these substances particularly in connection with the sister technique of classical total intensity light scattering (now commonly referred to as "static" light scattering to distinguish it from "dynamic" light scattering).

The decline of sedimentation equilibrium and other ultracentrifuge techniques in the 1970s and 80s in the protein biochemistry field (largely because of the advent of electrophoretic procedures and gel permeation chromatography) and followed by the revival of interest has been well documented (see e.g., Ref. [38]). In the polysaccharide and synthetic polymer fields the decline of sedimentation equilibrium as a routine absolute molecular weight tool has been largely because of the advent of laser light scattering techniques (both "classical" and "dynamic"). Probably the bulk of the apparatus and the length of time required to reach equilibrium have contributed to its downfall, but in reality light scattering – although the apparatus is more compact and measurements themselves are a lot quicker – suffer far worse disadvantages [39], largely through sample clarification, and for this reason it is fair to say that light scattering results have a greater degree of uncertainty than sedimentation equilibrium, when used in isolation. Although sedimentation methods would be my own method of

choice, we find particularly for molecular weight distribution work confirmatory measurements from both techniques (and used in conjunction with gel permeation chromatography) extremely valuable (see, e.g., Ref. [3]). We place greater value on literature light – scattering data that has been supported by independent sedimentation analysis than otherwise.

References

1. Harding SE (1992) In: Harding SE, Rowe AJ, Horton JC (eds) Analytical Ultracentrifugation in Biochemistry and Polymer Science Chap 27, Royal Society of Chemistry, Cambridge, UK
2. Harding SE (1989) Adv Carbohyd Chem & Biochem 47:345–381
3. Harding SE, Berth G, Ball A, Mitchell JR, Garcia de la Torre J (1991) Carbohyd Polym 16:1–15
4. Giebler R (1992) In: Harding SE, Rowe AJ, Horton, JC (eds) Analytical Ultracentrifugation in Biochemistry and Polymer Science Chap 2, Royal Society of Chemistry, Cambridge, UK
5. Yphantis DA (1964) Biochemistry, 3:297–317
6. Harding SE (1984) Biochem J 219:1061–1064
7. Fujita H (1975) Foundations of Ultracentrifuge Analysis, Chap 5, Wiley and Sons, New York
8. Creeth JM, Pain RH (1967) Prog Biophys Mol Biol 17:217–287
9. Creeth JM, Harding SE (1982) J Biochem Biophys Meth 7:25–34
10. Harding SE, Horton JC, Morgan PJ (1992) In: Harding SE, Rowe AJ, Horton JC (eds) Analytical Ultracentrifugation in Biochemistry and Polymer Science Chap 15, Royal Society of Chemistry, Cambridge, UK
11. Laue TM (1992) In: Harding SE, Rowe AJ, Horton JC (eds) Analytical Ultracentrifugation in Biochemistry and Polymer Science Chap 6, Royal Society of Chemistry, Cambridge, UK
12. Rowe AJ, Wynne-Jones S, Thomas DG, Harding SE (1992) In: Harding SE, Rowe AJ, Horton JC (eds) Analytical Ultracentrifugation in Biochemistry and Polymer Science Chap 5, Royal Society of Chemistry, Cambridge, UK
13. Teller DC (1965) PhD Dissertation, University of California, Berkeley, California, and (1973) Meth Enzymol 27:346–441
14. Harding SE (1985) Biophys J 47:247–250
15. Gilbert GA, Gilbert LM (1980) J Mol Biol 144:405–408
16. Creeth JM, Harding SE (1982) Biochem J 205:639–641
17. Harding SE, Johnson P (1985) Biochem J 231:549–555
18. Kawahawa K, Ohta K, Miyamoto H, Nakamura S (1984) Carbohyd Polym 4:335
19. Sato T, Norisuye T, Fujita H (1984) Macromolecules 17:2696
20. Woodward JR, Phillips DR, Fincher GB (1983) Carbohyd Polym 1983, 3:143
21. Errington N, Harding SE, Vårum KM, Illum L (1993) Int J Biol Macromol 15:113–117

22. Edmond E, Farquhar S, Dunstone JR, Ogston AG (1968) Biochem J 108:755
23. Muzzarelli RAA, Lough C, Emanuelli M (1987) Carbohyd Res 164:433
24. Harding SE, Rowe AJ, Creeth JM (1983) Biochem J 209:893–896
25. Berth G, Dautzenberg H, Lexow D, Rother G (1990) Carbohyd Polym 12:39
26. Lecacheux D, Mustiere Y, Panaras R, Brigand G (1986) Carbohyd Polym 6:477
27. Wedlock DJ, Baruddin BA, Phillips GO (1986) Int J Biol Macromol 8:57
28. Horton JC, Harding SE, Mitchell JR, Morton-Holmes DF (1991) Food Hydrocolloids, 5:125–127
29. Jumel K, Mitchell JR, Harding SE (1993) in preparation
30. Creeth JM, Cooper B (1984) Biochem Soc Trans 12:618–621
31. Mannion RO, Melia CD, Launay B, Cuvelier G, Hill SE, Harding SE, Mitchell JR (1992) Carbohyd Polym 19:91–97
32. Harding SE, Creeth JM (1982) IRCS (Int Res Commun System) Med Sci Lib Compend 10:474–475
33. Roark D, Yphantis DA (1969) Ann NY Acad Sci 164:245–278
34. Howlett GL, Jeffrey PD, Nichol LW (1972) J Phys Chem 76:77
35. Herdan G (1949) Nature 163:139
36. Lechner MD (1992) In: Harding SE, Rowe AJ, Horton JC (eds) Analytical Ultracentrifugation in Biochemistry and Polymer Science Chap 16, Royal Society of Chemistry, Cambridge, UK
37. Harding SE, Ball A, Mitchell JR (1988) Int J Biol Macromol 10:259–264
38. Schachman HK (1992) In: Harding SE, Rowe AJ, Horton JC (eds) Analytical Ultracentrifugation in Biochemistry and Polymer Science Chap 1, Royal Society of Chemistry, Cambridge, UK
39. Harding SE (1988) Gums & Stabilisers for the Food Industry 4:15–23

Received July 5, 1993;
accepted September 23, 1993

Authors' address:

Dr. Stephen E. Harding
University of Nottingham
School of Agriculture
Sutton Bonington LE12 5RD, United Kingdom

Progress in Colloid & Polymer Science   Progr Colloid Polym Sci 94: 66–73 (1994)

Sedimentation analysis of potential interactions between mucins and a putative bioadhesive polymer

I. Fiebrig[1,2]), S. E. Harding[1]), and S. S. Davis[2])

University of Nottingham,
[1]) National Centre for Macromolecular Hydrodynamics, School of Agriculture, Sutton Bonington and
[2]) Department of Pharmaceutical Sciences, University Park, Nottingham, England

Abstract. A potentially fruitful but hitherto relatively unexplored application of the analytical ultracentrifuge is the investigation of interactions in pharmaceutical polymer drug carrier systems. In this study we investigate the interactions of the cationic material chitosan with gastric mucin: 1) The physicochemical properties of gastric mucin and chitosan (highly deacetylated chitosan "Sea Cure + 210") in solution are summarised; 2) We described how the gastric mucin (from pig) can be purified by a combination of preparative isopycnic density gradient ultracentrifugation and gel chromatography, and how its purity and structural integrity can be checked by analytical isopycnic density gradient ultracentrifugation (on a MSE Centriscan 75 Analytical Ultracentrifuge) and on line GPC/MALLS (gel permeation chromatography coupled on-line to a multi-angle laser light scattering photometer); 3) Using co-sedimentation experiments with the appropriate controls in an XL-A Ultracentrifuge (absorption optics) and an MSE Mk II Analytical Ultracentrifuge (Schlieren optics) a definite interaction between chitosan and pig gastric mucin was demonstrated, with complexes sedimenting faster than 1000 S, depending on the amount of the mucin used.

Key words: Drug delivery – bioadhesion – chitosan – mucin – analytical ultracentrifugation

Introduction

I. Sedimentation velocity and bioadhesion

Sedimentation analysis provides a potentially powerful tool for the investigation of many phenomena relevant to pharmaceutical sciences. One such application so far relatively unexplored is the evaluation of putative bioadhesive or mucoadhesive polymers as drug carriers for oral drug delivery systems [1]. The performance of such a mucoadhesive polymer in vivo will be dictated by its ability to "adhere" in a controllable way to the mucus lining of the stomach or small intestine, thus *increasing* the transit time of a dosage form and hence *enhancing* the drug absorption from the intestine [2].

Although there are many factors which can influence any possible interaction phenomena in vivo (such as pH, characteristics of the mucus layer, intestinal contents, motility), a fundamental study on potential interaction phenomena between a mucin (the polyanionic glycoprotein component of mucus which dictates its characteristic physical properties of high viscosity and viscoelasticity [3]) and the putative mucoadhesive in *dilute solution* is a required baseline for understanding the mechanism of interaction. However, it is appreciated that solute concentrations and environmental conditions can be quite different to the in vivo situation and the relevance of such basic physicochemical studies has yet to be ascertained.

For dilute solution interaction phenomena, the principle of *co-sedimentation* of component macromolecules in a mixture is a particularly valuable one for assessing the strength of an interaction, and this principle is the cornerstone of our study here, in which we investigate the behaviour of mixtures of mucins with the cationic biopolymer chitosan; the latter molecule being of interest because of its opposite charge to the mucin [4]. Unfortunately small intestinal mucin is very difficult to harvest in amounts sufficient for a thorough investigation, so in this pilot study we investigate only the behaviour of mixtures of pig gastric mucin (as a model for human gastrointestinal mucin) and chitosan. The limitations of using pig gastric mucin (which has a lower charge density as opposed to small intestinal mucin in pigs as well as humans [5]) have to be taken into account.

In this study we describe the basic solution properties of pig gastric mucin and chitosan and their admixture. We also describe the use of analytical density gradient sedimentation equilibrium to assess purity and GPC/MALLS (gel permeation chromatography on line with multi-angle laser light scattering) to assess structural integrity in terms of distribution of molecular weight of pig gastric mucin. The degree of interaction between pig gastric mucin and chitosan in dilute solution (*without* chromophore labelling) has been measured by comparing sedimentation rates of mucin absorbance boundaries scanned with UV absorption optics in the Optima XLA Ultracentrifuge and by comparing residual areas under Schlieren sedimentation diagrams (recorded in the MSE MkII analytical ultracentrifuge) for the chitosan in the presence and absence of mucin.

II. Mucus and mucins

The mucus layer which protects the underlying epithelium of the gastrointestinal tract, is composed, apart from ~ 95% water, mainly of mucus glycoprotein or "mucin" [6, 7]. It is the mucin component of mucus which dictates its physical properties (such as high viscosity, viscoelasticity and *spinnbarkeit*) and which might play an important role in bioadhesive drug delivery. Mucins from a wide variety of sources (besides gastrointestinal, for example, bronchial, cervical and ovarian) consist of a similar building block or "basic unit" (Fig. 1). This basic unit consists of a polypeptide backbone which is heavily glycosylated and has a molecular weight of

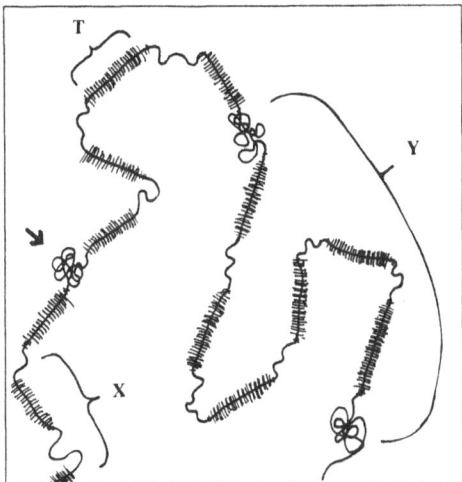

Fig. 1. Linear model for cervical mucin consisting of *basic units*: "T-domains" linked by naked peptide regions, some linked by disulfide bridging [16] (Arrowed region is susceptible to thiol attack) Key: T, "T-domain,"; X, "basic unit" and Y, "subunit"

~ 50 0000 g/mol. The carbohydrate side chains are built from 5–30 residues each [8–10]. The basic units are linked via naked (i.e. carbohydrate free) end regions into linear arrays, giving structures whose weight average molecular weights can range from 1×10^6 (e.g. ovarian cyst mucin) to over 40×10^6 (see, e.g. [11]). Thiol reduction gives rise to "subunits" consisting of 3–4 basic units. It is important to note that many of the carbohydrate chains contain sialic acid as their terminal sugar [12]. This rather acidic sugar is derived from neuraminic acid and has a pKa of 2.6. Fully dissociated under most physiological pH conditions, it gives the mucin molecule a net negative charge, allowing the possibility of ionic interaction with positively charged polymers.

III. Chitosan

In this study, chitosan, a derivative of chitin, was chosen as a possible mucoadhesive. It is particularly attractive because of its polycationic properties. It is composed of linear chains of *N*-acetyl glucosamine units with different degrees of deacetylation, upon which the charge density of a given chitosan preparation will depend [13] (Fig. 2).

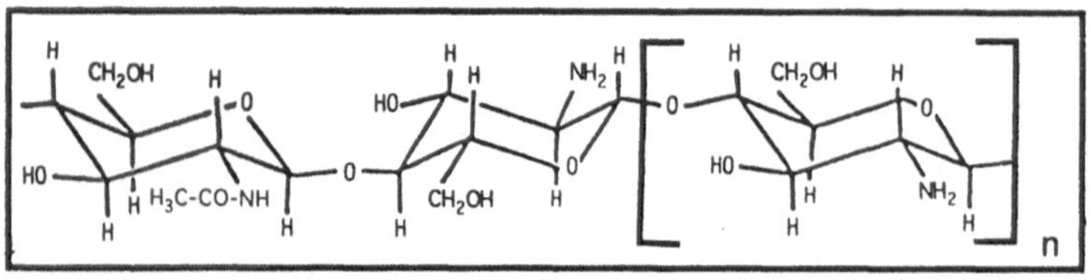

Fig. 2. Structure of chitosan, a derivative of chitin with different degrees of deacetylation of the basic *N*-acetyl glucosamine unit

Table 1. Characterization of Chitosan Sea Cure + 210 (SC + 210) [10]

Property		Units	Values
$\bar{v}$	partial specific volume	(mg/g)	0.580($\pm$ 0.011)
DA	degree of acetylation	(wt%)	11
M_w	weight average molecular weight	(g/mol)	162($\pm$ 10)$\times 10^3$
B	second virial coefficient	(ml mol g^{-2})	2.75 $\times 10^{-2}$
BM_w		(ml g^{-1})	4455
$s^0_{20,w}$	sedimentation coefficient at zero concentration	(S)	1.41($\pm$ 0.05)
K_s	sedimentation coefficient concentration regression coefficient	(ml/g)	88.6($\pm$ 10.8)
$[\eta]$	intrinsic viscosity	(ml/g)	540($\pm$ 20)
$Dz_{20,w}$	translational diffusion coefficient	(cm^{2}s^{-1})	3.9($\pm$ 0.6)$\times 10^{-9}$

Materials

Solutions

For all sedimentation velocity analyses an acetate buffer pH 4.0 I = 0.1 [14] was used. For the analytical isopycnic density gradient experiments on the mucin, a loading density of $\rho_e = 1.346$ g/ml was achieved by the addition of an appropriate amount of Cs$_2$SO$_4$ (analytical grade) to a mucin solution at a nominal concentration of 3 mg/ml.

Mucin

Fresh pig gastric mucus glycoprotein (PGM) was purified from fresh solubilized pig gastric mucus by preparative caesium chloride isopycnic density gradient ultracentrifugation in an enzyme inhibitor cocktail according to a modified procedure of Hutton et al. [15]. This was followed by gel permeation chromatography of the glycoprotein fraction on a Sepharose CL-2B column. The totally excluded volumes were pooled and concentrated by ultrafiltration, dialyzed against distilled water and fractions of 1ml solution kept frozen at $-20\,°$C or freeze dried. The mucin preparation was gently defrosted and dialysed into the buffer or redissolved in buffer before use.

Chitosan

Sea Cure + 210, a glutamate salt of chitosan (Protan Ltd., Drammen, Norway) was used. This was an 11% acetylated preparation which had been previously well characterised in this laboratory. Table 1 summarises the relevant physical properties from earlier work performed by our group [16].

Experimental

Mucin purity-analytical isopycnic density gradient ultracentrifugation

This method was used to assess the purity of the pig gastric mucin following the procedures of Creeth et al. [17]. An MSE Centriscan 75 Analytical Ultracentrifuge was used, equipped with 280 nm absorption optics. The sample cell was loaded with the mucin/Cs$_2$SO$_4$ solution and run at a rotor speed of 50 000 rpm. A density gradient is built up by the caesium salt as a result of the centrifugal force, showing a lower density at the meniscus of the fluid column and an increasingly higher density towards the bottom of the cell. After reaching equilibrium approximately 48 h later, the glycoprotein fraction with a buoyant density of approximately 1.33 g/ml will concentrate at the centre of the fluid column, free protein will be

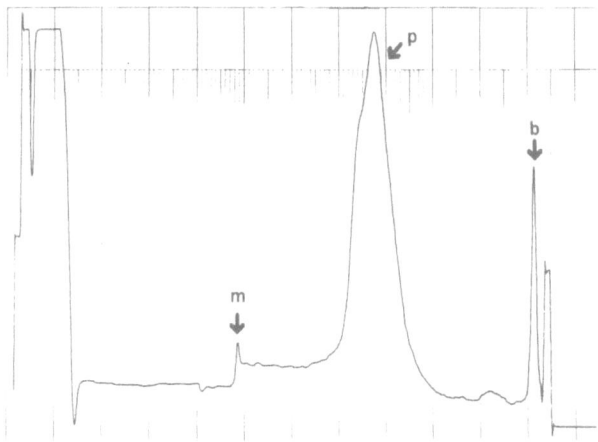

Fig. 3. Pig gastric mucus glycoprotein (PGM) in a density gradient of Cs_2SO_4 run on a MSE Centriscan 75 Analytical Ultracentrifuge at a rotor speed of 50000 rpm. Loading density $\rho_e = 1.346$g/ml. Key: m = solution meniscus, ρ = glycoprotein peak [calculated density at peak: $\rho_p = 1.337$g/ml ; b = cell base. The small shoulder on the lower density flank of the peak is due to density heterogeneity of the material which is not unusual for glycoprotein preparations of this kind [23, 24]

'floating' at the meniscus with a buoyant density of 1.24 g/ml [18] and any nucleic acid will be concentrated at the cell bottom. The densities at each point of the fluid column can be calculated by the equations given by Creeth and Horton [19]. The mucin appears as one clear peak at a calculated density of 1.337 g/ml in accordance with published values by [20]. There was no free protein nor nucleic acid detectable (Fig. 3).

Mucin Purity and Structural Integrity – GPC/MALLS ~ (gel permeation chromatography on line with multi angle laser light scattering)

As a relatively rapid assay for the structural integrity of the mucin, a Dawn-F Gel Permeation Chromatography/Multi Angle Laser Light Scattering (GPC/MALLS) system was used [21]. For the GPC, two columns (packing material: Hydroxyethylmethacrylate crosslinked with ethylenglycol dimethacrylate) in series were used: one PSS Hema Bio linear followed by a PSS Hema Bio 40 (Mainz, FRG). The manufacturer's specified separation range of the column system for dextran was from *above* 1×10^6 to *below* 5000. Samples were injected at a nominal loading concentration of 1mg/ml. The high dilution during elution through the scattering cell (by approximately 10x) meant that correction for non-ideality effects was not necessary. After passing through the scattering cell, solute concentrations were recorded on-line by a Waters 410 differential refractive index detector. The MALLS chromatogram (Fig. 4) shows a very pure mucin material (symmetric peak), free of low molecular weight impurities.

Interaction studies–mucin detection

In the first set of experiments of mixture of defrosted solution of PGM (at a concentration of ~ 0.4 mg/ml and a weight average molecular weight of $M_W \approx (9.3 \pm 0.2) \times 10^6$) and a solution of SC + 210 (at a concentration of ~ 2 mg/ml) was run on a Beckman Optima XL-A Analytical

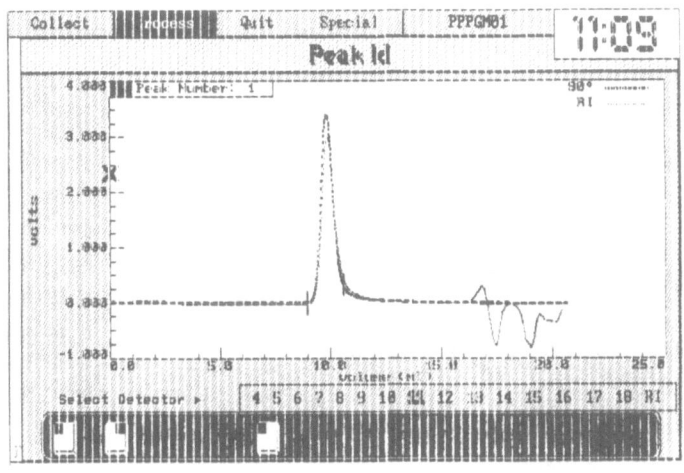

Fig. 4. Pig gastric mucus glycoprotein (PGM): Chromatogram from a Dawn-F Gel Permeation Chromatography/Multi-Angle Laser Light Scattering (GPC/MALLS) system. The trace of the refractive index detector overlays closely with the light scattering signal, the sample appears to be very pure and fairly homogeneous i.e. with no degraded mucin present. The noise on the far righthand side of the refractive index trace is due to small amounts of buffer salt eluting

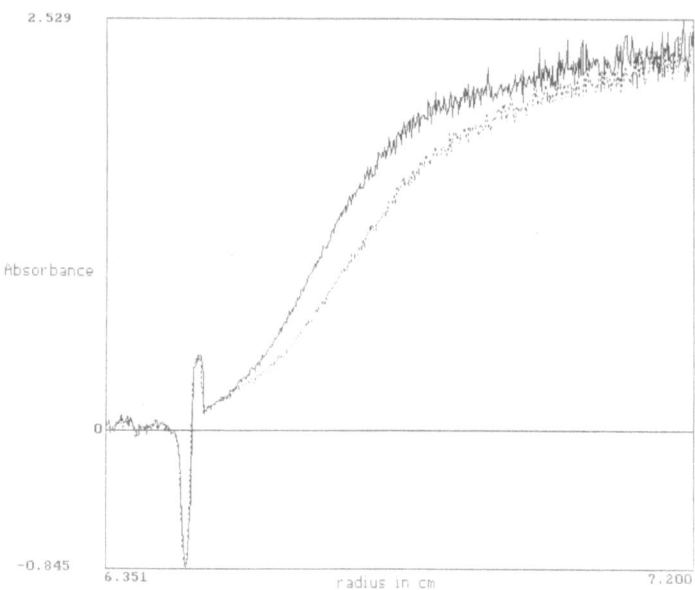

Fig. 5. PGM/Sea Cure + 210-complex in mixture cell, X-LA Analytical Ultracentrifuge, rotor speed 2000 rpm, scan interval 10 min, absorption optics at $\lambda = 230$ nm, sedimentation coefficient for complex: (1990.0 ± 18.0)S

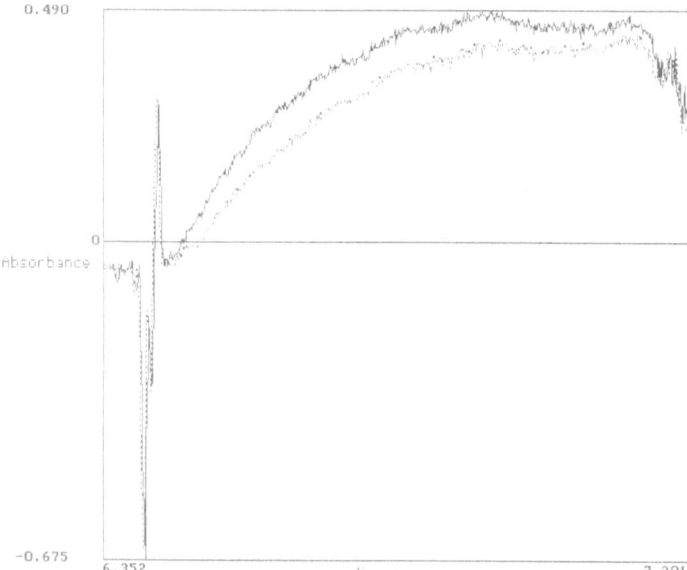

Fig. 6. PGM control cell, X-LA Analytical Ultracentrifuge, rotor speed 10 000 rpm, scan interval 6 min, absorption optics at $\lambda = 230$ nm, sedimentation coefficient for mucin: (53.0 ± 2.8) S. Due to heterogeneity of the material the mucin boundary is quite shallow, very unlike the sharp sigmoidal boundaries typical for many proteins

Ultracentrifuge (equipped with UV/Vis absorption optics) against a mucin control diluted to an equal concentration. This centrifugation technique was used to follow the mucin and mucin/chitosan boundaries. Cells (12 mm pathlength) in a four-hole rotor were employed and the samples run at 37 °C at a rotor speed of 2000 rpm first (Fig. 5), to detect the sedimenting boundary of the mucin/chitosan mixture (advantage was taken here of the stability of the XL-A even at low rotor speeds), then speeded up to 10 000 rpm to detect the fast moving boundary of the mucin control (Fig. 6). Sedimentation velocity traces were analysed on a digitizing pad connected to a computer equipped with software to generate sedimentation coefficients at concentrations corrected for radial dilution [22]. The value for the mixture cell was of (1990.0 ± 18.0) S, as opposed to a sedimentation coefficient of $(53.0 \pm 2.8) =$ S for the mucin control.

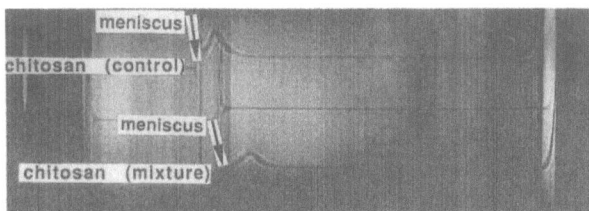
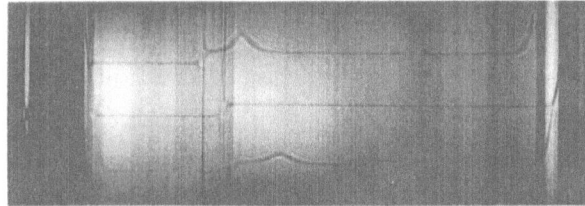
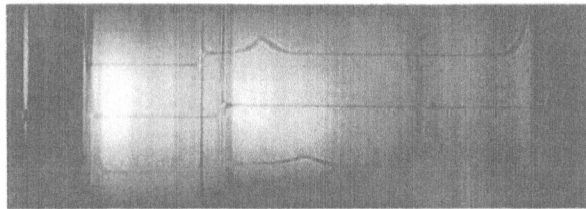
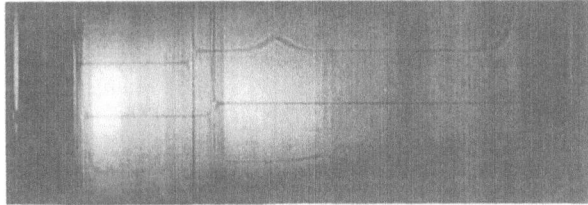

Fig. 7. Chitosan control [(2.1 $\pm$ 0.03) S] as well as residual chitosan in mixture cell [(2.5 $\pm$ 0.02) S], MSE Mk II Analytical Ultracentrifuge, rotor speed 35 000 rpm, Schlieren optics

Interaction studies–chitosan detection

Because the chitosan was "invisible" for UV detection at the wavelength used in the XLA studies above, it was important to detect the position and rate of movement of the chitosan boundary in the control, as well as to establish how much chitosan has been lost through the high molecular weight complex formation in the mixture. A required higher concentration of mucin was achieved by dissolving $\sim$ 3 mg/ml of the freeze-dried PGM sample in buffer. Three 20 mm pathlength cells in a rotor were loaded onto the MSE Mk II Analytical Ultracentrifuge (equipped with conventional Schlieren optics): One cell containing a 1:1 mixture of 4 mg/ml SC + 210 solution and $\sim$ 3 mg/ml mucin solution, weight average molecular weight of $M_W \approx$

(6.1 $\pm$ 0.2) $\times 10^6$. The other one containing the chitosan control and the third one containing the mucin control. The sedimentation velocity traces were analysed as mentioned above. Schlieren optics detected the sedimenting mucin control boundary which sedimented at (53.2 $\pm$ 1.0) S at a rotor speed of 15 000 and a temperature of 36 °C. The chitosan control sedimented at (2.1 $\pm$ 0.03) S. The mixture (mucin-chitosan complex sedimented too fast to be detected) left a residual amount of chitosan behind sedimenting at (2.5 $\pm$ 0.02) S (Fig. 7). The mixture was later run separately on the XL-A Ultracentrifuge at 37 °C, showing a sedimentation coefficient of (10 300.0 $\pm$ 250.0) S for the complex. This compares with the value of (1990.0 $\pm$ 18.0) S referred to above, which corresponds to approximately seven fold less mucin, indicating that the loading concentration of mucin in the mixture is a critical factor in governing the size of the complex. Due to the limited amount of highly purified mucin material further experiments on concentration dependence of the complex size were not possible. All given sedimentation coefficients were not corrected to standard conditions–differences are considered to be minor compared to the large changes observed through complex formation. The corrected sedimentation coefficient ($s_{20,w}^0$) for our chitosan control is in good agreement with the corresponding value given in Table 1.

Discussion

A clear interaction between isolated and purified pig gastric mucin (major component of gastric mucus) and a polycationic derivative of chitin (as potential bioadhesive drug carrier) was found, evidenced by a significant increase of the sedimentation coefficient of the mucin in combination with chitosan. The mucin control sedimented at (53.2 $\pm$ 1.0) S [$M_W \approx$ (6.1 $\pm$ 0.2) $\times 10^6$], 1.5 mg/ml, and (53.0 $\pm$ 2.8) S [$M_W \approx$ (9.3 $\pm$ 0.2) $\times 10^6$], 0.2 mg/ml. The chitosan control sedimented at (2.1 $\pm$ 0.03) S. For the mixture we obtained values of 10 300.0 $\pm$ 250.0) S and (1990 $\pm$ 18.0) S respectively, for mucin concentrations of 3.0 mg/ml and 0.4 mg/ml.

As expected in an excess of chitosan the mucin/chitosan complex sediments first, leaving residual chitosan behind. Comparisons of the areas under the curve, between the chitosan control (AUC: chitosan = 1) and the residual chitosan in the mixture cell (AUC: chitosan in mixture = 0.69) shows by subtraction an approximate chitosan/mucin ratio of 1 : 2 (w/w ratio) and 3.75 : 1 (molecular ratio). The increased sedimentation coefficient of chiotsan in the mixture [(2.5 ± 0.02) S compared with the chitosan control [(2.1 ± 0.03) S] is probably due to a lower viscosity of the solution as a result of the decreased concentration of chitosan. The sensitivity of the size of the complex to the amount of mucin could have implications for the in vivo situation, where the mucin concentrations (in gel/sol form) are likely to be much higher.

In conclusion, chitosan would appear to have some potential as a bioadhesive. However, we have only examined mucin/chitosan interactions at very low concentrations and at a single pH. In vivo not only water and mucin, but also proteins, lipids etc. may be present in the mucus layer and pH values can vary from as low as pH 1.5 in the stomach up to over pH 7.5 in the small intestine. These results nevertheless should form the firm basis of experiments more relevant to in vivo conditions. As to the actual mechanisms of the interaction, this will require further detailed study such as the effects of pH, ionic strength and other solvent conditions (e.g. the presence or absence of bile salts) and this will be subject of a further study.

Finally this is yet another example of how a combination of sedimentation methods can be used together to assay for an interaction between molecules at high dilution.

Acknowledgements

We would like to thank Prof. A. Allen and F. Fogg from the Department of Physioloical Sciences, Newcastle, for their valuable help and information on the preparation of mucin as well as K. Jumel for the GPC/MALLS analyses and P. Mistry for technical assistance. The University of Nottingham would like to thank Hoechst U.K.Ltd for their financial support.

References

1. Anderson MT (1991) The Interaction of Mucous Glycoproteins with Polymeric Materials, Ph.D. thesis, University of Nottingham, England
2. Robinson JR (1990) Rationale of bioadhesion/mucoadhesion. In: Gurny R, Junginger HE (eds) Bioadhesion: Possibilities and Future Trends. Wissenschaftliche Verlagsgesellschaft mbH, Stuttgart, pp 13–15
3. Allen A (1981) Structure and function of gastrointestinal mucus. In: Johnson LR (ed) Physiology of the Gastrointestinal Tract by Raven Press, NY, pp 617–639
4. Lehr C-M, Bouwstra JA, Schacht EH, Junginger HE, (1992) In vitro evaluation of mucoadhesive properties of chitosan and some other natural polymers. Int. J. Pharmaceutics 78:43–48
5. Allen A (1989) Gastrointestinal mucus. In: Schultz SG, Forte JG, Rauner BB (eds) Physiology of the Gastrointestinal Tract – The GI-System III, American Physiological Society, Bethesda, Maryland, pp 359–382
6. Carstedt I and Sheehan JK (1988) Structure and macromolecular properties of mucus glycoproteins. Monogr. Allergy. Karger Basel, 24:16–24
7. Neutra MR and Forstner JF (1987) Gastrointestinal mucus: Synthesis, secretion and function. In: Johnson LR (ed) Physiology of the Gastrointestinal Tract, 2nd ed. Raven Press, NY, pp 975–1009
8. Silberberg A (1987) A model for mucus glycoprotein structure Biorheology 24:605–614
9. Carlstedt I, Sheehan JK, Corfield AP and Gallagher JT (1985) Mucus glycoproteins. In: A Gel of a Problem, Essays in Biochemistry 20:40–76
10. Allen A (1978) Structure of gastrointestinal mucus glycoprotein and the viscous and gel forming properties of mucus. Br Med Bull 34:28–33
11. Harding SE (1989) The macrostructure of mucus glycoproteins in solution. Adv Carb Chem Biochem 47:345–383
12. Schauer R. (1992) Sialinsäurereiche Schleime als bioaktive Schmierstoffe. Nachr Chem Tech Lab 40, Nr 11:1227–1231
13. Sandford, PA (1989) Chitosan: Commercial uses and potential applications. In: Skjåk-Braek G, Anthonsen T, Sandford P (eds) Chitin & Chitosan, pp 51–69
14. Dawson MC, Elliot DC, Elliot WH, Jones KM (1986) Data for Biochemical Research, Clarendon Press Oxford, p. 429
15. Hutton DA, Pearson JP, Allen A, Foster SNE (1990) Mucolysis of the colonic mucus barrier by faecal proteinases: Inhibition by interacting polyacrylate. Clinical Science 78:265–271
16. Errington N, Harding SE, Vårum KM, Illum L (1993) Hydrodynamic characterization of chitosans varying in degree of acetylation. Int J Biol Macromol 15:113–117
17. Creeth JM, Bhaskar KR, Horton JR (1977) The separation and characterization of bronchial glycoproteins by density-gradient methods. Biochem J 167:557–569
18. Ifft JB, Vinograd J (1966) The buoyant behaviour of bovine serum mercaptalbumin in salt solutions at equilibrium in the untracentrifuge. II. Net hydration ion binding and solvated molecular weight in various salt solutions. J Phys Chem 70:2814–2822
19. Creeth JM, Horton JR (1977) Macromolecular distribution near the limits of density-gradient columns, Biochem J 161:449–463
20. Creeth JM, Denborough MA (1970) The use of equilibrium-density-gradient methods for the preparation and

characterization of blood-group-specific glycoproteins. Biochem J 117:879–891

21. Wyatt PJ (1992) Combined differential light scattering with various liquid chromatography separation techniques In: Harding SE, Sattelle DB, Bloomfield VA (eds) Laser Light Scattering in Biochemistry. The Royal Society of Chemistry, London, pp 35–58

22. King DJ, Byron OD, Mountain A, Waeir N, Harvey A, Iawson ADG, Proudfoot KA, Baldock D, Harding SE, Yarranton GT, Owens RJ (1993) Expressions, purification and characterization of B72.3 Fv fragments. Biochem J 290:723–729

23. Creeth JM, Bhaskar KR, Horton JR (1977) The separation and characterization of bronchial glycoproteins by density gradient methods. Biochem J 167:557–569

24. Harding SE, Creeth JM (1983) Polyelectrolyte behaviour in mucus glycoproteins. Biochim et Biophys Acta 746:114–119

Received June 18, 1993;
accepted October 13, 1993

Authors' address:

Mr. Immo Fiebrig
Department of Pharmaceutical Sciences
University Park
Nottingham NG7 2RD, United Kingdom

Progress in Colloid & Polymer Science   Progr Colloid Polym Sci 94:74–81 (1994)

An on-line interferometer for the XL-A ultracentrifuge

T. M. Laue, A. L. Anderson, and P. D. Demaine

University of New Hampshire, Department of Biochemistry and Molecular Biology, Spaulding Life Sciences Building, Durham, New Hampshire, USA

Abstract: As a result of the renewed interest in analytical ultracentrifugation, the Beckman XL-A was released recently. This instrument automates spectrophotometric measurements of the concentration distribution of molecules in a gravitational field. Presented here is the design and the performance characteristics of a Rayleigh interferometer for the XL-A ultracentrifuge. The interferometer consists of a laser diode light source, imaging optics providing a 1.4-fold magnification of the cell, and a solid-state television camera detector. The source and detector are interfaced to a personal computer which synchronizes data acquisition for up to four cells, performs data reduction, and allows data analysis. About 1700 data points, at a radial spacing of $\sim 9~\mu m$, are acquired over a standard double sector cell image. Data are then stored on the disk and presented as a graph on the screen. The complete sequence of data acquisition, storage, and presentation requires about 15 s. The interferometer has a precision of about ± 0.003 fringe, and can be used for both equilibrium and velocity sedimentation. The design can be adapted for Schlieren detection.

Key words: Analytical ultracentrifuge – interferometry – instrumentation – equilibrium sedimentation – velocity sedimentation – methods – hydrodynamics – thermodynamics

Introduction

The principle measurement needed for any analytical ultracentrifuge experiment is the solute concentration as a function of radial position. As currently configured, the Beckman XL-A contains a superb spectrophotometer. However, it is desirable to complement this optical system with refractometric detection. The Rayleigh interferometer provides a cell image in which the refractive index difference between sample and reference at each radial position is given by the vertical displacement of a set of evenly spaced horizontal fringes [1, 2]. This image is well-suited for automated data acquisition [3–6].

Absorbance and interference (refractive) optical systems provide complementary ways to determine concentration distributions in the centrifuge. The merits of the absorbance system are that it can provide sensitive and selective solute detection. For many biochemical systems these virtues are needed and appreciated. However, it is less useful when solutes do not absorb significantly or when solvents do. For these cases refractive detection offers several advantages over absorbance measurements. Moreover, the interferometer provides greater accuracy [7], higher radial resolution [8, 9], a greater concentration range, and the ability to trace very steep concentration gradients [8].

Presented here is the description of a Rayleigh interferometer suitable for use with the Beckman XL-A analytical ultracentrifuge. This instrument is suitable for use in both sedimentation equilibrium [8–10] and sedimentation velocity analysis [11].

Description of the on-line interferometer hardware

Overview

A schematic of the on-line Rayleigh interferometer is presented in Fig. 1. The two periscopes

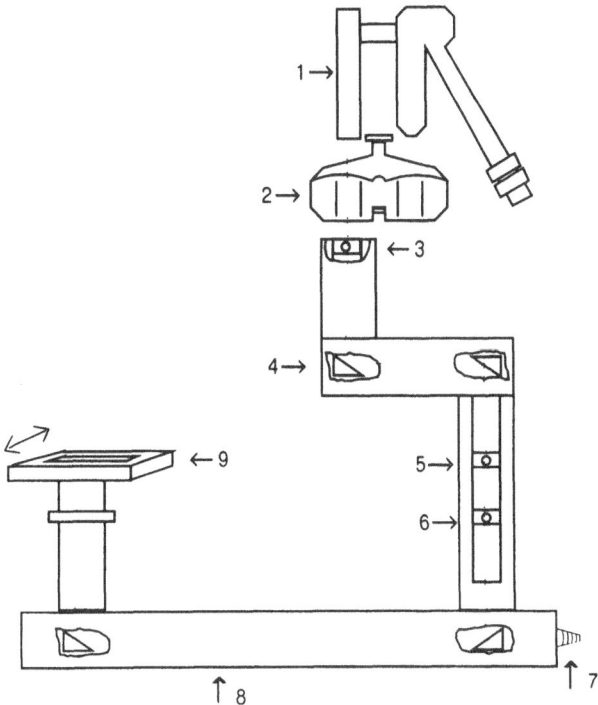

Fig. 1. Overview of the XL-A interference optical system. 1) light source, 2) rotor, 3) condensing lens, 4) top periscope, 5) cylinder lens, 6) camera lens, 7) vacuum port, 8) bottom periscope, and 9) television camera sensor. The heat sink of the XL-A was machined to accept the condensing lens holder and top periscope. For safety reasons, two separate vacuum systems are used, one for the rotor chamber and a second one to evacuate the optical track

& Hoyer Microbench (Göttingen, FRG) system were used. The optics consist of a pulsed laser diode light source, beam expansion and collimation optics, slits to provide the reference and sample beams, the centrifuge cell, a condensing lens to mix the two beams, a cylinder lens and a camera lens to focus the cell image and fringes, respectively, on the CCD television camera. The entire optical track, except for the television camera, can be kept under a vacuum. This eliminates distortion of the image due to the Schlieren effects caused by convective air currents. Both the laser and the camera are operated by computer so that their functions can be synchronized to one another and to the spinning rotor. The computer also is used to acquire images, calculate the concentration distribution, store information, and to edit and analyze the data.

Lightsource

The light source consists of a Toshiba TOLD 9215 10 mwatt, 670 nm, index guided laser (Fig. 2). A pulsed, constant-current circuit was developed for driving the laser (Fig. 3). Two lenses, a 16-mm focal length achromat mounted approximately 5–8 mm from the laser diode, and an 80 mm achromat provide a collimated beam of light sufficiently large to illuminate the Rayleigh mask. The custom Rayleigh mask, two 0.5 mm slits, 2.5 cm long and separated by 0.8 cm (kindly supplied by Beckman Instruments) is mounted in a rotatable holder in front of the 80 mm achromat lens. The light source mount is attached to the XL-A monochromator arm, and both optical

are necessary to avoid obstacles and to fold the optical path to fit inside the machine. This design is based on prototypes constructed using the Model E ultracentrifuge [10]. Wherever possible, stock optical components from the Spindler

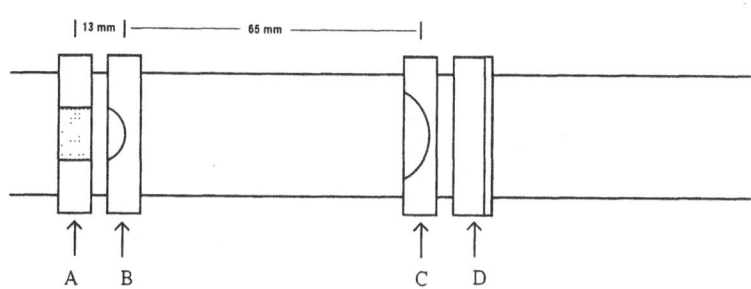

| 13 mm | ———— 65 mm ———— |

A B C D

Fig. 2. Schematic of the XL-A light source. Lenses were chosen to expand and collimate the laser beam. The resulting beam is a $1.2 \, cm \times 3 \, cm$ ellipse. Distances shown are nominal, with the final positions chosen to provide the best collimation, as monitored by a shear-plate [10]. Components include: A) the 10 m watt, 670 nm, index-guided laser diode mounted in a Delrin holder for electrical isolation, B) a 16-mm fl 6 mm ϕ achromat, C) an 80-mm 31.5 ϕ achromat and D) Rayleigh mask holder. All components, except the Rayleigh mask, are from Spindler & Hoyer, and are mounted on 150-mm-long Microbench stainless steel rods cut to the appropriate length

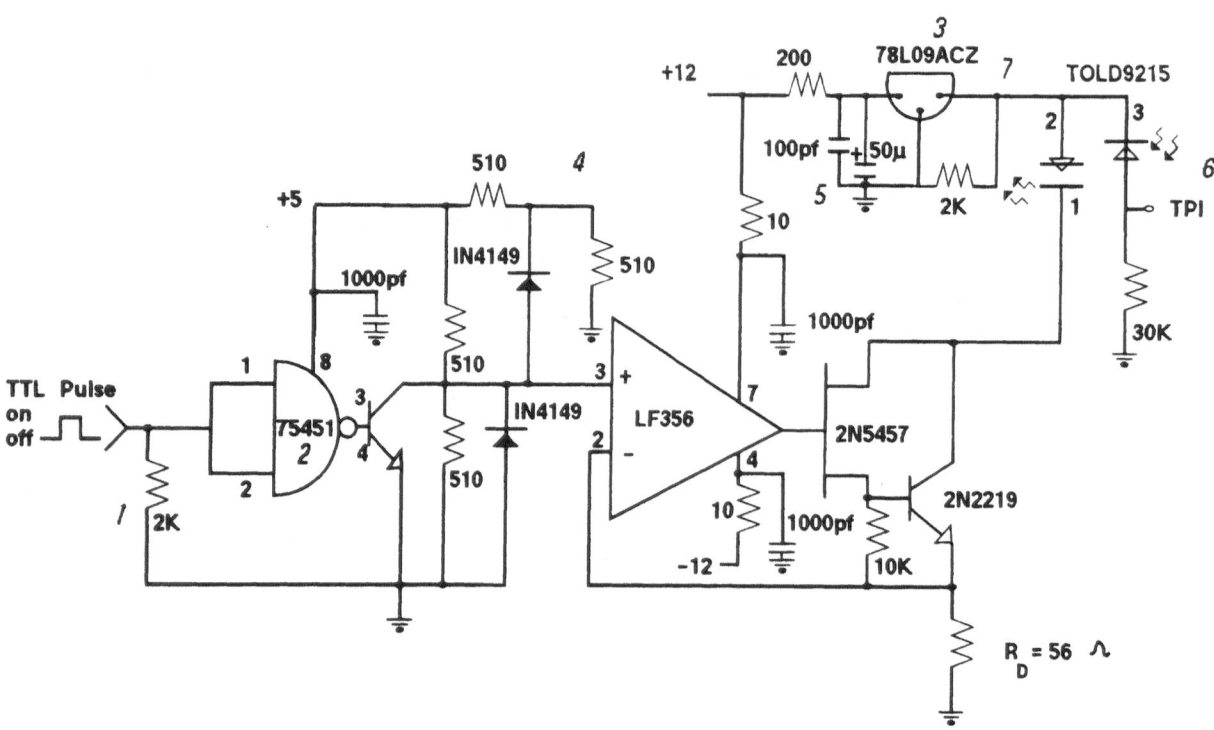

Fig. 3. Schematic of the XL-A light source pulsed constant-current power supply. A standard design for a constant current power supply has been adapted for pulse operation. The current out is equal to the input voltage (nominally a TTL signal) divided by R_D (here, 56 Ω). Considerable isolation and surge protection are provided to prevent diode failure. The power supply is attached to the light source mounting to minimize the distance between the power supply output and the laser diode. This means the power supply is in the rotor chamber, so appropriate materials and components should be chosen for vacuum compatibility. All power is derived from the $\pm$ 12 V supplies of the XL-A, including the 5 V for the 75451 (obtained using a 78L05ACZ, not shown) and the 9 V for the laser. The TTL-compatible pulse signal is brought in on shielded cable using pin 21 of the vacuum feed through in the rotor chamber base plate. The two 510 Ω and two 1N4149 diodes clamp the output of the 75451 to values between 0.3 and 3.0 volts. A built-in photodiode provides a signal proportional to the laser power (TP1). The laser rise time measured at TP1 $\approx$ 500 ns and the fall time is only slightly longer. Because of the slight offset bias on the LF356, the laser "sinters" (i.e., is always slightly forward biased) which helps reduce the pulse rise time. R_D determines the current passing through the laser diode, and should be chosen to keep the current within the laser specification when the TLL input is fully high (5 volts). As shown here, the current is switched from about 5 ma to 50 ma when the input voltage exceeds 2.1 volts. This power supply should be able to operate lasers with drive currents up to 100 ma

systems may be operated simultaneously. Through careful machining, only three adjustments are needed, two for pointing the laser beam and one for rotating the slits. The beam pointing adjustments are arranged to be parallel and perpendicular to the rotor radius. A small hole (1 mm) between the two slits provides a collimated pencil of light that is used during alignment, but is blocked during normal operation. The slits are rotated to be parallel to the rotor radius [1, 2]. All of these adjustments only need to be made once.

Synchronizing the laser to the spinning rotor

Pulsed lasers permit the use of multicell rotors without using wedged-windows, and provide a superior image [12]. However, their use requires that timing circuits be used to ensure that the light pulses consistently illuminate the same arc of the rotor. In this case, a counter-timer computer interface (CTM-05, MetraByte, Taunton, MA, USA) serves as a rate-multiplier circuit [4]. A rotor timing pulse is derived from the Hall-effect sensor built into the XL-A. The period of one

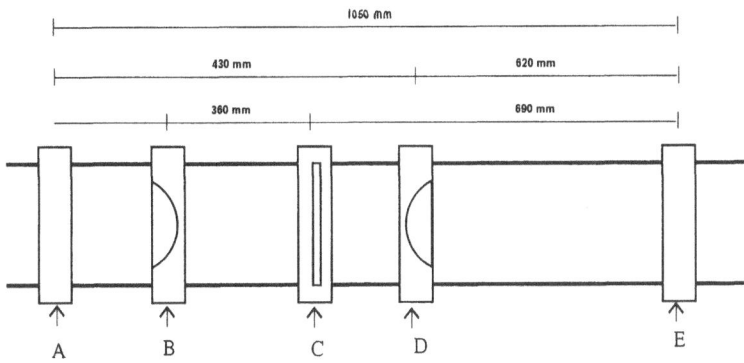

Fig. 4. Schematic of the XL-A Rayleigh imaging optics. The object plane (A) is roughly 50 mm from the 250-mm fl achromat condensing lens (B). The condensing lens also separates the rotor chamber vacuum system from the optical track vacuum system. The 63-mm fl cylinder lens (C) is in a custom mount that allows it to be rotated and translated (see text). The 250-mm fl achromat camera lens (D) is at a fixed position on the optical track, and focusing is achieved by shifting the camera plane (E) along its mounting post

revolution is determined by counting a 6 MHz clock between rotor timing pulses. The value in the counter is multiplied by a fraction corresponding to the portion of a revolution to wait before pulsing the laser. A second counter is loaded with the resulting value then decremented, starting at the next rotor timing pulse, by the same clock used to determine the rotor period. When the count equals zero, the laser is pulsed. The counter is updated every second allowing rapid rotor acceleration to be tracked. The software includes algorithms to take account of the rotor-speed dependence of the laser delay [4]. The duration of the laser pulse is determined in the same fashion, so that a constant position and constant arc are illuminated. If the pulse is of very short duration, then the Rayleigh slits on the cell may be eliminated, thus facilitating the simultaneous use of the scanner and interference optical systems.

Imaging optics

The basic optical design (Fig. 1) is the same as that described previously [1, 2], except that the overall optical path is shortened by approximately half to 1000 mm. The imaging optics are presented in a schematic form in Fig. 4. The 250-mm focal length condensing lens is mounted on the heatsink of the XL-A and provides the vacuum seal. This lens is approximately 50 mm from the 2/3 position in the centrifuge cell, considered to be the image plane [1, 2, 8]. The light from the condensing lens is passed through a periscope to bring it out from the crowded center of the XL-A and into the optical bench. A beam steering mirror at the end of the first periscope centers the optical beam on the optical track. A 63-mm focal length cylindrical lens is placed so that its focal

point coincides with the focal point of the condensing lens [1, 2]. The cylinder lens is mounted in a custom-built holder that permits the lens to be rotated and translated. Rotation is used to align the cylinder lens axis with the slits in the Rayleigh mask. Translation of the lens is perpendicular to the cylinder lens axis, and is used to center the interference image on the television camera. A 250-mm focal length achromat serves as the camera lens and is positioned about 400 mm from the image plane. This provides a nominal 1.4-fold magnification of the image at the plane of the sensor. A second periscope is used to fold the optics so that they fit in the housing of the XL-A. A beam steering mirror at the end of the optical track ensures that the light beam is centered on the television camera. The television sensor is attached to a vertical tube (Fig. 1), permitting it to be moved for focusing and allowing the camera to be rotated so that rows of pixels are parallel to the cell radius.

Solid state camera

A television camera replaces photography in the on-line interferometer. In order to keep costs reasonable, a standard format (RS-170 or PAL), high-resolution CCD camera is used. The available cameras are satisfactory, but not ideal – the major drawback being the sensor size. Conformance to the 2/3-inch camera tube format standard means that the sensors are nominally 6–8 mm wide by 4–6 mm high with about 600–750 pixels (depending on the manufacturer) in each row. Since it is desirable to attain a radial resolution of 10 μm or better, the cell image must be magnified, making it too large to fit on the sensor. For this reason, the television camera is moved past the magnified image, even though this results in greater hardware and software complexity.

For the system described here a Philips VCM3250 (Philips Industrial Devices, Inc. Eindhoven, The Netherlands) imaging module was used. This device provides an RS-170 signal containing information from 512 pixels per row and 492 pixels per column, with a pixel pitch of 10 μm horizontal and 12.8 μm vertical. The camera is attached to a movable stage made from a stepping motor and linear stage from the head actuator of a standard 5-1/4-inch computer floppy disk drive, yielding 40 camera positions across the cell image. Modifications to the disk drive include the removal of all extraneous components and hardware, then machining the disk-drive mounting plate to fit under the XL-A. An adapter was machined to attach the plate to the vertical tube of the optical track.

The operation of the television camera, video signal digitizing and temporary data storage are provided by a "frame grabber" interface. Suitable interfaces are available from many companies. The system described here uses a Data Translation DT2851 interface (Data Translation, Inc., Marlborough, MA, USA) in an IBM-PC compatible computer. This interface converts the RS-170 composite analog video signal into a 512 horizontal by 480 vertical array of 8-bit numbers. Using composite video can lead to the limitations described previously [10]. Undoubtedly, an optimal system would achieve high radial resolution with minimum distortion by using a larger array, slow-scan operation of the camera and an appropriate interface to insure a one-to-one correspondence between the pixels and the resulting data array. To date, the added expense of such a system has not been justified.

Computer system

Operation of the Rayleigh optical system is optimized so that the application interface can be kept as simple as possible for general operation, while providing more experienced users access to the full capabilities of the machinery. This means, of course, that the behind-the-scenes work of the operating system is fairly complicated. The prototype system uses an 8-Mbyte, 33 MHz, 80486DX-based IBM-PC compatible computer for all operations, including synchronizing the

light source, acquiring the data, converting the image to fringe displacement as a function of radial position, and data storage. Since there is a desire to edit and analyze data during the course of an experiment, it is necessary for the acquisition software to be multitasked with other software. For this reason, all code for running the real-time interferometer is written in C++ operating under Microsoft Windows. All calibration factors and operating preferences are stored in an initialization file. This eliminates the need for any further calibration by the user. Even so, all calibrations are automated to simplify machine setup or optical changes.

Alignment and calibration

The alignment and focusing of the Rayleigh system is performed essentially as described previously [2, 4]. Some modifications to these procedures ease the alignment and take advantage of the new optics and the real-time display of the image. A paper detailing the alignment procedure is in preparation.

Once the camera is aligned, each column of pixels on the sensor corresponds to a unique radial position. Two magnification factors, along with the positions of the two reference positions provided by the rotor counterweight, relate the position of the motorized stage and the pixel column index to the radial position. The midpoint between the reference edges is 6.5 cm from the center of rotation, so that the radial position can be calculated from any steeper position and column index as:

$$r = 6.5 + E \cdot M_{\mathrm{m}}^{-1} + (256 - \mathrm{i}_a)$$
$$\cdot M_{\mathrm{c}}^{-1} \left[\frac{\mathrm{ref}_a + \mathrm{ref}_b}{2} \right], \tag{1}$$

where ref_a and ref_b are the positions corresponding to the inner and outer reference edges, E is the motorized stage position in steps (i.e., 0–39), M_{m}^{-1} is the inverse of the magnification factor for the motorized stage, i_a is the column index, and M_{c}^{-1} is the inverse of the camera magnification factor [10, 13].

Fringe displacement measurement

In the Rayleigh interferometer, the fringes are equally spaced at the image plane, regardless of the refractive index difference between the solvent and solution or the gradient in refractive index [1, 3, 8]. The relative vertical displacement of the fringes is obtained as the phase of a single-frequency, discrete Fourier analysis conducted at the spatial frequency of the fringes, as described previously [10, 13]. Software is provided that automates the determination of the correct Fourier frequency. The Fourier method yields the fractional fringe displacement, which is converted to the full fringe displacement as described previously [13]. The maximum traceable gradient is about 50 fringes/mm [13], which is ample for all but the steepest gradients, where limitations in accuracy due to Wiener skewing become a concern [1, 8].

Imperfections in the optics are corrected by subtracting a blank scan taken of just the Rayleigh optics (i.e., without a cell positioned in the beam). This is achieved by timing the laser pulse so that a scan is taken in one of the indentations on the rotor. The blank scan is acquired, stored and used without user intervention. This feature may be shut off by the user, if so desired.

Even after careful alignment, the cell image may need to be rotated slightly [4, 8, 10]. Image rotation is adjusted automatically using the slope of a best-fit line to the blank image, using the algorithm described previously [10]. No user intervention is needed, though image rotation may be shut off, if desired.

Data presentation and storage

The sequence of data acquisition, graphic presentation, and data storage requires about 15 s. The Origin (MicroCal, Inc., Marlborough, MA) graphics engine is used for data presentation. Options are provided that allow the user to select the axes. Y-axis options are either fringe displacement or the time derivative of the fringe displacement. The latter is useful for both equilibrium analysis, where a horizontal line indicates equilibrium has been reached, and for sedimentation velocity, where boundaries are converted to peaks [11]. The X-axis can either be the radius or the apparent sedimentation coefficient [11]. With these options, it is possible to have a real-time display of the progress of a sedimentation velocity experiment.

Data are stored as ASCII files using the same format as that currently used for absorbance data

on the XL-A. The first column is the radial position, r in cm, the second is the fringe displacement, $Y(r)$ in units of fringes, and the third is the Fourier magnitude at that point. The latter value is useful in editing data since points with a low magnitude correspond to places where the image quality was poor. Data are stored in this format regardless of how the user chooses to display the data.

System performance

From the standpoint of speed, the performance of the on-line system is superb compared to manual data acquisition, which required hours. However, the more important issue is the accuracy of the concentration distribution. To this end, only two issues must be addressed: 1) the precision of the fringe displacement measurement, and 2) the accuracy of the radial position measurement.

The precision of the fringe displacement measurement can be addressed in several ways [10]. One test of the precision is to examine the reproducibility of fringe displacement measurements using real samples. Preliminary results indicate that the interferometer has a reproducibility of about ± 0.002 fringe. It is hard to judge the accuracy of the interferometer, though our experience with curve fitting suggests that ± 0.01 fringe is a conservative estimate.

The accuracy of the radial position determination is affected by the accuracy of: 1) the motorized linear stage, 2) the accuracy of the pixel positioning on the sensor, 3) the accuracy of the two magnification factors, and 4) the accuracy of the reference edge positions. Of these elements, the error in the magnification factors seem to be the most serious [10]. Calculation based on the estimated error in these values leads to a maximum estimated error of ± 12 µm at either edge of the image and less error towards the center of the image.

Prospects and conclusion

In order for analytical centrifugation to regain its place as a useful method that can be conducted rapidly and with certainty, the automation of the data acquisition is essential. The on-line Rayleigh interferometer presented here fulfills this requirement. However, the extreme rapidity with which precise data can be acquired extends analytical ultracentrifugation in two other ways. First, it makes possible the accomplishment of far more complicated experimental protocols than could be considered when photographs were used. For example, it is reasonable to determine the molecular weight of a dozen or more samples in less than an hour using short column methods [9] and the on-line interferometer. This permits a rapid means of surveying buffer conditions, performing titrations and testing sample homogeneity. The on-line system is useful for examining unstable compounds, where often it is possible to monitor sample stability during the course of an experiment. Likewise, it is now a relatively simple prospect to gather the large quantity of high precision data necessary to characterize interacting systems fully. Finally, it has proven to be indispensable for the investigation of heterogeneous associations, where a large number of conditions need to be examined to assure the accuracy of the model.

The second way the on-line interferometer extends analytical ultracentrifugation is by permitting the development of entirely new methods. For example, the ability to acquire fringe displacement measurements in a few seconds allows, for the first time, the rapid determination of $dY(r)/dt$. By transforming the radial axis to the apparent sedimentation coefficient, it is possible to calculate the differential sedimentation coefficient distribution function on-line. Figure 5 presents sedimentation velocity data from the XL-A interferometer, and shows how such graphs may be used diagnostically.

It is quite likely that other methods will be developed that rely on the capabilities of the on-line interferometer. These new methods should extend the utility of analytical ultracentrifugation both in the industrial and the academic sectors. New uses will include rapid and certain quality control for pharmaceuticals, polymers and dispersions, as well as new insights into the realms of macromolecular interactions. For example, by combining the data from the interferometer and the spectrophotometer, it is possible to determine the extinction coefficient for a material using as little as 15 μl of sample. Non-constancy of the extinction coefficient across a boundary can provide a rapid diagnostic for the heterogeneity of the

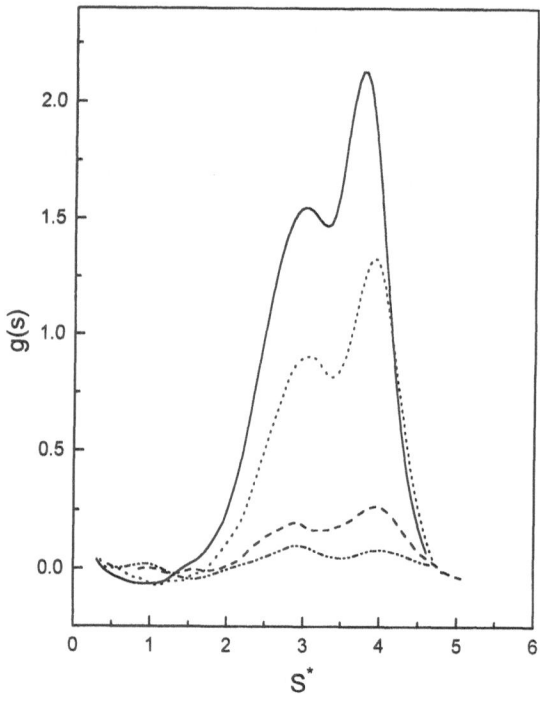

Fig. 5. Concentration dependence of the derivative sedimentation coefficient distribution function, $g(s)$, as a function of the apparent sedimentation coefficient, S^*, for bovine skin decorin in 20 mM Tris, pH 8.0, 150 mM NaCl, 5 mM EDTA and 5 mM EGTA. Data were acquired at 50 000 rpm, 20°C, and at protein concentrations of 0.13 (dot-dash), 0.25 (dash), 1.0 (dot) and 1.5 (solid) mg/ml. The increase in material at higher S^* with increasing protein concentration is diagnostic for a mass-action reversible self-association. A sequence of 10 scans, taken at 2-min intervals and 75 min after acceleration, were averaged to produce these results [11]

material. Certainly, the instrument is capable of these functions, and we must wait to see what other new applications are developed.

There are certain applications which are best served by Schlieren optics. Straightforward adaptation of the design presented here should be possible. Changes that would need to be made include: 1) placement of a rotatable analyzer plate at the focal plane of the condensing lens, 2) modification of the laser source to act as a line, rather than as a point source, and 3) changes to the software to account for the resultant image [14]. With this addition, the XL-A will have the full complement of optical systems presently available for the Model E.

Acknowledgments

This work was supported by grants BBS 86–15815 and DIR 90–02027 from the National Science Foundation. The author thanks David Yphantis for our many years of work together, Theresa Ridgeway for help with the figure and critical reading of the manuscript, Jun Liu for providing the sedimentation velocity data, and Beckman Instruments for supplying the XL-A, parts, advice and encouragement during this development. This is scientific contribution No. 1840 from the New Hampshire Agricultural Experiment Station.

References

1. Richards EG, Schachman HK (1959) J Phys Chem 63:1578–1591
2. Richards EG, Teller DC, Schachman HK (1971) Anal Biochem 41:189–214
3. DeRosier DJ, Munk P, Cox DJ (1972) Anal Biochem 50:139–153
4. Laue TM "Rapid Precision Interferometry for the Analytical Ultracentrifuge," Ph.D. Dissertation, Univ of Connecticut, Storrs, CT, 1981
5. Teller DC (1967) Anal Biochem 19:256–264
6. Richards JH, Richards EG (1974) Anal Biochem 62:523–530
7. Perlman GE, Longsworth LG (1948) J Am Chem Soc 70:2719–2724
8. Yphantis DA (1964) Biochemistry 3:297–317
9. Yphantis DA (1960) Ann New York Acad Sci 88: 586–601
10. Laue TM (1992) In: Harding SE, Rowe A, Horton JC (eds) Analytical Ultracentrifugation in Biochemistry and Polymer Chemistry, Royal Chemical Society, Cambridge, pp 63–89
11. Stafford WF III (1992) In: Harding SE, Rowe A, Horton JC (eds) Analytical Ultracentrifugation in Biochemistry and Polymer Chemistry, Royal Chemical Society, Cambridge, pp 359–393
12. Yphantis DA (1979) In: Hirs, CHW, Timasheff, SN (eds) Methods in Enzymology Volume 61, Academic Press, New York, pp 3–12
13. Laue TM (1993) In: Cohn G (ed), Ultrasensitive Clinical Laboratory Diagnostics, SPIE Proceedings, Vol 1895, SPIE, Bellingham, Washington, pp 18–27
14. Rowe AJ, Wynne Jones S, Thomas DG, Harding SE (1992) In: Harding SE, Rowe A, Horton JC (eds) Analytical Ultracentrifugation in Biochemistry and Polymer Chemistry, Royal Chemical Society, Cambridge, pp 49–62

Received August 19, 1993;
accepted September 23, 1993

Authors' address:

Dr. Thomas M. Laue
University of New Hampshire
Department of Biochemistry and Molecular Biology
Spaulding Life Sciences Building
Durham, New Hampshire, 03824. USA

Progress in Colloid & Polymer Science Progr Colloid Polym Sci 94:82–89 (1994)

The sedimentation diffusion equilibrium of a ternary gel

W. Borchard

Angewandte Physikalische Chemie der Universität GH Duisburg, FRG

Abstract: The well known theory of a binary gel, which is considered to be an elastic liquid mixture consisting of a crosslinked polymer component and a low molecular solvent, is extended to a ternary system, where instead of the pure solvent a polymer solution is given. The equations describing the continuous equilibria are derived for the solution and the gel phase. As soon as the gel is placed into the centrifugal field, an osmotically active pressure in the gel phase appears, which is identical to a swelling pressure. The latter can be calculated from the distribution of the solvent and the soluble uncrosslinked component in the gel phase. At higher fields a discontinuous phase boundary gel/solution will occur. In case no gel phase is present, the system reduces to a binary solution in the centrifugal field. At the phase boundary gel/solution the well known deswelling effect is expected if the soluble polymer is expelled from the network. The discussion includes the necessary experimental quantities for the determination of the thermodynamic properties of physically crosslinked gels where a soluble part is present which is not built into the network but is acting as a second solvent.

Key words: Ternary gel – continuous equilibrium – swelling – swelling pressure – centrifugal field

1. Introduction

In recent years the deformation of a binary elastic fluid mixture, called a gel, in a centrifugal field has been treated in a general way [1]. The gel consisting of a crosslinked polymer and a low molecular compound has been assumed to remain isotropic in the deformed state. During the first stages of the action of the external centrifugal field a concentration gradient inside the gel will establish, which leads to an increase of the polymer concentration in the direction of the radial distance from the axis of rotation. From the law of mass conservation it follows that the polymer concentration is diminished at the gel surface in contact with the vapor. But in contrast to solutions, the polymer concentration at this phase boundary cannot go gradually to zero, because the lowest polymer concentration has to correspond to the concentration of the swollen cross-linked system in the heterogeneous swelling equilibrium. This is the reason why this phase boundary of the gel starts to sediment as soon as the polymer concentration has reached the

value of the swelling equilibrium. With increasing polymer concentration at the cell bottom the mass conservation of the crosslinked polymer causes the gel/vapor surface to move in the direction of increasing radial distance r [2]. These findings are in agreement with early observations of Svedberg and Pedersen, who stated that the movement of the gel surface, which was defined as sedimentation of the gel, did not occur in all cases [3]. It is clear that a partial sedimentation of the crosslinked polymer takes place as soon as a concentration gradient of the crosslinked polymer starts to establish. There is no need of defining an incubation period, because the continuous equilibrium in the gel phase may establish without any movement of the gel surface, if we disregard compression and mixing effects.

From these considerations it follows directly that the gel surface has to move, if a gel swollen to maximum is placed into the centrifugal cell, and if the external field is acting.

As, in general, the molar mass of a gel is extremely high, where sometimes the total polymer mass belongs to a huge crosslinked

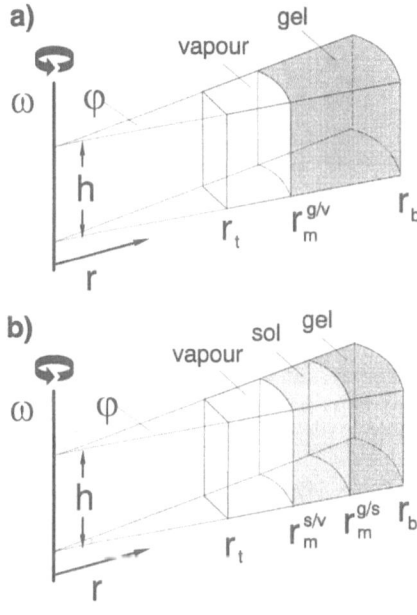

Fig. 1. Schematic representation of a sector shaped analytical cell of an ultracentrifuge filled with gel; a) at time $t = 0$ and b) at sedimentation equilibrium $t \to \infty$, r = radial distance from the axis of rotation, ω = angular velocity, φ = sector angle of the centerpiece, h = height of the centerpiece, indices: t = top of the cell, m = meniscus, b = bottom of the cell.
Phase boundaries: g/v = gel/vapor, s/v = solution/vapor, g/s = gel/solution (see text)

macromolecule, we expect a gel to change its concentration in a centrifugal field much quicker than in the case of a polymer solution with the same type of polymer but much lower molar mass. Thus, on movement of a binary gel surface the pure solvent or, in the case of a ternary gel, a polymer solution will be left, leading finally to the situation which is schematically described in Fig. 1. It is supposed that a gel is placed into a centrifugal cell with an angle of the sector ϕ and height h at time $t = 0$ (Fig. 1a). From this time onward the cell is rotating with an angular velocity ω. The gel is situated between the bottom of the cell r_b and the meniscus between gel (g) and vapor (v), $r_m^{g/v}$, where its surface is in contact with the vapor.

The vapor extends to the radial distance $r = r_t$, which corresponds to the top of the cell. At sufficiently high angular velocities the gel is supposed to move towards the bottom of the cell, which is schematically shown in Fig. 1b. After a long enough time, there are three phases in a continu-

ous equilibrium with the phase boundaries between gel and solvent or solution, $r_m^{g/s}$, and between solvent or solution and the vapor, $r_m^{s/v}$, where m indicates the meniscus.

We are mainly interested in the formulation of the case that the gel consists of three components which all may achieve the continuous equilibrium by free movements due to sedimentation and diffusion forces. This includes that the gel is in contact with a solution. Therefore, we expect that the already treated binary case is a special case of the ternary gel [1].

2. Theory

A ternary gel consists of three components i. The low molecular solvent is characterized by $i = 1$, the crosslinked polymer by the index 2, and the oligomer or second solvent by the index $i = 3$. The solution or solvent phase is indicated by the phase index ', the gel by the index ''. Because of the small vapor pressure of the solvents the influence of the vapor phase is neglected.

The specific driving force in an external field, $\tilde{X}_i$, consists of two parts, namely, the negative gradients of the specific potential of i in the external field, $\tilde{\phi}_i$, and of the specific chemical potential of i, $\tilde{\mu}_i$ [4, 5]. The relation is presented in Eq. (1), where in the rotating system with a single axis of rotation only the radial distance from this axis r is of importance

$$\tilde{X}_i^\alpha = - \frac{d\tilde{\phi}_i^\alpha}{dr} - \frac{d\tilde{\mu}_i^\alpha}{dr} . \tag{1}$$

For $\alpha = ''$, we have $i = 1, 2, 3$. In the solution, which is given by $\alpha = '$, we have only components $i = 1, 3$, because the crosslinked polymer is only present in the gel phase.

If at time $t = 0$ the gel is brought into the centrifugal field, the driving force $\tilde{X}_i$ is caused only by the first term on the righthand side of Eq. (1). At later stages the changes of the concentrations i in the gel will lead to increasing gradients of the specific chemical potential of component i and finally to the sedimentation diffusion equilibrium, in which the driving forces $\tilde{X}_i$ have to vanish.

For a centrifugal field we have:

$$-\frac{d\tilde{\phi}_i^\alpha}{dr} = \omega^2 r , \tag{2}$$

where i is used in the same way as described above. This means that the negative gradient of i only depends on the angular velocity ω and the radial distance r, where the righthand side of Eq. (2) is the centrifugal acceleration. If the sedimentation takes place in a gravitational field, this term has to be substituted by the gravitational acceleration.

It was mentioned that the gels are supposed to be isotropic, which means that only the pressure dependence of $\tilde{\mu}_i$ has to be considered instead of the different tensor components of the stresses and deformations. Additionally, the independent concentration variables in both phases have to be introduced. If the mass concentration of component i is used, there are only two independent concentration variables in the gel phase, where ρ_1'' and ρ_3'' are taken. In the sol phase we may choose either ρ_1'' or ρ_3''. Thus, we get for the solution, taking ρ_1' as the independent concentration variable:

$$\frac{d\tilde{\mu}_i'}{dr} = \tilde{V}_i' \frac{dP'}{dr} + \left(\frac{\partial \tilde{\mu}_i'}{\partial \rho_1'}\right)_{T,P} \frac{d\rho_1'}{dr} ; \quad i = 1 \text{ or } 3 , \tag{3a}$$

and for the gel

$$\frac{d\tilde{\mu}_i''}{dr} = \tilde{V}_i'' \frac{dP''}{dr} + \left(\frac{\partial \tilde{\mu}_i''}{\partial \rho_1''}\right)_{T,P} \frac{d\rho_1''}{dr} + \left(\frac{\partial \tilde{\mu}_i''}{\partial \rho_3''}\right)_{T,P}$$

$$\times \frac{d\rho_3''}{dr} ; \quad i = 1, 2, 3 , \tag{3b}$$

where $\tilde{V}_i^\alpha$ is the partial specific volumes of i in phase α and P the pressure. Using the mechanical equilibrium condition in a continuous system [4, 5], which is given by

$$\rho^\alpha \omega^2 r = \frac{dP^\alpha}{dr} ,$$

where $\rho^\alpha = \sum_i \rho_i^\alpha$ and $\sum_i \rho_i^\alpha \tilde{V}_i^\alpha = 1$ in both phases, we obtain by multiplying Eq. (1) by ρ_i^α and summation over all components in each phase by use

of Eqs. (2, 3).

$$\sum_{i=1,3} \rho_i'\left(\frac{\partial \tilde{\mu}_i'}{\partial \rho_j'}\right)_{T,P} \frac{d\rho_j'}{dr} = 0 ; \quad j = 1 \text{ or } 3 \tag{5a}$$

for the solution phase, and

$$\sum_{i=1}^{3} \rho_i''\left(\frac{\partial \tilde{\mu}_i''}{\partial \rho_j''}\right)_{T,P} \frac{d\rho_j''}{dr} = 0 ; \quad j = 1, 3 \tag{5b}$$

for the gel phase. These Gibbs-Duhem equations combine the gradients of the different specific chemical potential of i in the different phases.

Finally, we arrive by use of Eqs. (1–5) at expressions for the solvent phase

$$(1 - \tilde{V}_1' \rho') \omega^2 r \, dr = [d\tilde{\mu}_1']_{T,P} , \tag{6a}$$

or

$$(1 - \tilde{V}_3' \rho') \omega^2 r \, dr = [d\tilde{\mu}_3']_{T,P} . \tag{6b}$$

For the gel phase, we obtain three equations:

$$(1 - \tilde{V}_1'' \rho'') \omega^2 r \, dr = [d\tilde{\mu}_1''(\rho_1'', \rho_3'')]_{T,P} , \tag{7a}$$

$$(1 - \tilde{V}_2'' \rho'') \omega^2 r \, dr = [d\tilde{\mu}_2''(\rho_1'', \rho_3'')]_{T,P} , \tag{7b}$$

and

$$(1 - \tilde{V}_3'' \rho'') \omega^2 r \, dr = [d\tilde{\mu}_3''(\rho_1'', \rho_3'')]_{T,P} . \tag{7c}$$

In Eqs. (6a to 7c) the differentials of the chemical potentials have to be taken at constant T and P, where the starting or reference pressure for the solution is P_0 at the interface between vapor and solution and for the gel may be the pressure at the meniscus $r_m^{g/s}$ named P_{ref}. Thus, on integration of these equations only the different concentrations in both phases solution and gel have to be considered. Therefore, the integrals may be written in the abbreviated form for the solution:

$$\omega^2 \int_{r=r_m^{v/s}}^{r} (1 - \tilde{V}_i' \rho') r \, dr = [\Delta \tilde{\mu}_i'(\rho_j')]_{T,P} ;$$

$$i, j = 1 \text{ or } 3 , \tag{8a}$$

and for the gel:

$$\omega^2 \int_{r=r_m^{g/s}}^{r} (1 - \tilde{V}_i'' \rho'') r \, dr$$

$$= [\Delta \tilde{\mu}_i''(\rho_1'')]_{T,P} + [\Delta \tilde{\mu}_i''(\rho_3'')]_{T,P}; \quad i = 1, 2, 3 . \tag{8b}$$

If $\tilde{V}_i^\alpha$ and also ρ^α are independent of pressure P or distance r the phase α is considered to be incompressible.

The righthand sides of Eqs. (8a) and (8b) can be written in the form for the solution:

$$[\Delta \tilde{\mu}_i'(\rho_i')]_{T,P} = \tilde{\mu}_i'(T, P_0, \rho_i')$$
$$- \tilde{\mu}_i'(T, P_0, \rho_i^{v/s}) ; \quad i = 1, 3 ,$$
(9a)

and for the gel:

$$[\Delta \tilde{\mu}_i''(\rho_i'')]_{T,P} = \tilde{\mu}_i''(T, P_{\text{ref}}, \rho_i'')$$
$$- \tilde{\mu}_i''(T, P_{\text{ref}}, \rho_i^{g/s}) , \quad i = 1, 2, 3 .$$
(9b)

It can be seen that the difference of the chemical potentials of component i in Eq. (9a) is proportional to an osmotic pressure due to the difference in the partial density of $\rho_i' - \rho_i^{v/s}$. Due to Eq. (8a) it is zero, if $\omega^2 = 0$ or $\tilde{V}_i'\rho' = 1$. The difference is at maximum for the upper limit of the integral at $r = r_m^{g/s}$. Correspondingly, the difference of the chemical potentials of i in Eq. (9b) is proportional to a swelling pressure due to the difference in the partial densities $\rho_i'' - \rho_i^{g/s}$.

3. Discussion

3.1. The solution phase

The difference of the specific chemical potentials of component i in Eq. (9a) may be achieved by a differential osmotic pressure experiment. The oligomer, $i = 3$, is supposed to be hindered to pass the membrane, so that it is only permeable to the solvent. In this case, we may formulate

$$\tilde{\mu}_1'(T, P_0, \rho_1') - \tilde{\mu}_1'(T, P_0, \rho_1^{v/s})$$
$$= \Delta \tilde{\mu}_1'(T, P_0, \Delta \rho_1') \text{ with } \Delta \rho_1' \equiv \rho_1' - \rho_1^{v/s} ,$$
(10)

which is given by [6]

$$\Delta \tilde{\mu}_1'(T, P_0, \Delta \rho_1') = - \int_{P_0}^{P} \tilde{V}_1' \mathrm{d}P = - \int_{P_0}^{P_0 + \Pi_{\text{os}}} \tilde{V}_1' \mathrm{d}P .$$
(10a)

From these equations we read that we get a continuous change of $\tilde{\mu}_1'$ in a centrifugal experiment due to the concentration gradients, which

can be determined in a differential osmotic experiment only by defined steps of concentrations [7]. If the oligomer is polydisperse a separation of the components is obtained [5]. Therefore, different averages of the molar masses may be determined as has been practiced by Scholte [8]. The integral in Eq. (8a) may be solved if from the mass conservation equations there is an explicit expression for $\rho'(r)$ or $\tilde{V}_1'(r)$. If the rotational speeds are not too high, the solution may be approximately regarded as being incompressible, which means that $\tilde{V}_1'$ is independent of r.

A further limiting case may be given that at high rotational speeds the oligomer concentration is zero at $r = r_m^{v/s}$. This results in

$$\Delta \tilde{\mu}_1' = \tilde{\mu}_1'(T, P_0, \rho_1') - \tilde{\mu}_{01}'(T, P_0, \rho_{01}^{v/s}) .$$
(10b)

If the soluble polymer is of a very high molar mass that it does not enter the gel phase, we have at the phase boundary gel/solution

$$\Delta \tilde{\mu}_1' = \tilde{\mu}_1'(T, P_0, \rho_1^{g/s}) - \tilde{\mu}_{01}'(T, P_0, \rho_{01}^{v/s}) .$$
(10c)

We derive directly by means of Eqs. (1 and 2) the heterogeneous equilibrium condition in the differential form at the phase boundary between solution and gel

$$\mathrm{d}\tilde{\mu}_1' = \mathrm{d}\tilde{\mu}_1'' .$$
(10d)

Integrating both sides of Eq. (10d) starting from the pure solvent, we arrive at

$$\Delta \mu_1' = \Delta \mu_1'' ; \quad P = P_{\text{ref}} ,$$
(10e)

where at P_{ref} of the pure solvent $\tilde{\mu}_{01}'$ has to be that in the adjacent phase at the same pressure and temperature. Equation (10e) is the equivalent relation for the well known deswelling method of Boyer and Spencer, which has been widely used to determine the properties of swollen systems, namely, at pressure $P = P_0$ [9]. But Eq. (10e) may be easily calculated to the pressure P_0. From the last equation we derive that a gel, first contacted with the pure solvent and being at the solvent concentration of maximum swelling, has to deswell if it is contacted with a solution afterwards. This deswelling effect starts clearly from the swelling of the gel in the original concentrations of components 1 and 3 up to the swelling of the gel in the concentrations of components 1 and 3 at the boundary gel/solution in the sedimentation diffusion equilibrium. It can be imagined that the deswelling in a centrifugal field can be rather pronounced. Studying this effect in an ultracentrifuge is useful.

We now consider the case that the pure component 3, the oligomer chemically similar to the crosslinked polymer, is in the liquid state and occurs in phases ' and ". If no low molecular mass solvent is present, we read from Eq. (8a) that there is a complete equivalence for $i = 1$ or $i = 3$. This has to be the case, because now the pure component 3 is the only solvent for the gel. The ternary gel reduces to a binary gel, which has already been treated [1]. The special cases referring to Eqs. (10c–e) are not possible, because there are no equilibrium conditions, if neither component 2 is present in the solution phase, nor component 3 in the gel.

3.2. The ternary gel phase

The relations for a continuous equilibrium in a binary gel have been derived recently [1]. In the ternary gel three components, namely, the crosslinked polymer, the solvent, and the oligomer are present. In many practical cases component 3, the oligomer, is very similar to the crosslinked polymer in its chemical structure. Therefore the interaction energies between component i and j, namely, w_{ij}, are very similar. This will lead to $w_{12} \approx w_{13}$, $w_{22} \approx w_{33} \approx w_{23}$ which reduces the system to three independent quantities w_{11}, w_{12}, and w_{22} like in a binary polymer solution [10]. These hints are of importance, if the above derived continuous equilibria are simulated by simple relations from the statistical thermodynamic theory. This concerns as well the evaluation of the experimental results.

It is interesting that Eqs. (7a) and (7c) reduce to a completely symmetrical binary gel, if one of the two components 1 or 3 is missing:

$$(1 - \tilde{V}_1'' \rho'')\omega^2 r \, dr = [d\tilde{\mu}_1''(\rho_1'')]_{T,P} \quad \text{for } \rho_3'' = 0 , \tag{11a}$$

respectively,

$$(1 - \tilde{V}_3'' \rho'')\omega^2 r \, dr = [d\tilde{\mu}_3''(\rho_3'')]_{T,P} \quad \text{for } \rho_1'' = 0 . \tag{11b}$$

Equation (11b) is easily obtained from Eq. (11a) by the exchange of index 1 by index 3. The same result as in Eq. (11a) is obtained by changing the

index 3 by index 1 in Eq. (11b). Therefore, the ternary case is reduced in a quite general form to the binary case as can be expected. Eqs. (7a) to (7c) are consequently a complete description of the dependence of the specific chemical potentials i upon the concentrations of the independent variables ρ_1' and ρ_3''.

Equation (8b) delivers two contributions caused by the changes in concentration of components 1 and 3, both related to the pressure of reference P_{ref} at the boundary gel/solution by Eq. (9b) for $j = 1$ and $j = 3$. The difference of the chemical potential of component 1 in the gel

$$[\Delta \tilde{\mu}_1''(\rho_1'')]_{T,P,\rho_3''} = \tilde{\mu}_1''[T, P_{\text{ref}}, \rho_1''(r)]$$
$$- \tilde{\mu}_1''[T, P_{\text{ref}}, \rho_1''(r^{g/s})] \text{ for } \rho_3'' = \text{const} . \tag{9b}$$

is, in principle, a change which can be realized in a static experiment of a swelling pressure equilibrium. The difference in the concentrations of the solvent may be equalized as in a former contribution [1] by

$$[\Delta \tilde{\mu}_1''(\rho_1'')]_{T,P,\rho_3''} = - \int_{P_{\text{ref}}}^{P_{\text{ref}} + \pi_{s,1}} \tilde{V}_1'' \, dP , \tag{12a}$$

where $\pi_{s,1}$ is the swelling pressure of the gel due to the change of component 1. In the same way, we may define

$$[\Delta \tilde{\mu}_1''(\rho_3'')]_{T,P,\rho_1''} = - \int_{P_{\text{ref}}}^{P_{\text{ref}} + \pi_{s,2}} \tilde{V}_3'' \, dP \tag{12b}$$

for the oligomer with

$$[\Delta \tilde{\mu}_1''(\rho_3'')]_{T,P,\rho_1''} = \tilde{\mu}_1''[T, P_{\text{ref}}, \rho_3''(r)]$$
$$- \tilde{\mu}_1''[T, P_{\text{ref}}, \rho_3''(r^{g/s})] \quad \text{for } \rho_1'' = \text{const} . \tag{12c}$$

The introduction of the swelling pressures $\pi_{s,1}$ and $\pi_{s,2}$ is not necessary, because the changes of the specific chemical potentials of the solvent are determined by Eq. (8b). Therefore, the concentrations of components 1 and 3, the density of the gel and the partial specific volume of the solvent have to be known at every distance r. From the measurements of the refractive indices perhaps $\rho_2'' + \rho_3''$ are obtained. But for the evaluation of the equations ρ_3'' has to be determined separately. A further relation is derived from the mass conservation of the crosslinked polymer in the gel and the distribution of the solvent and the oligomer in the solution and the gel.

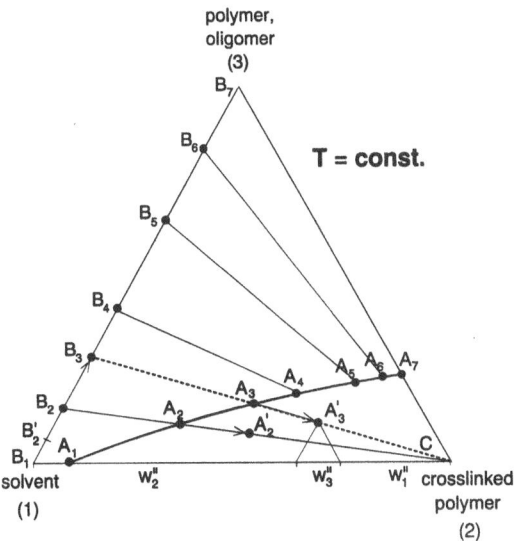

Fig. 2. Isothermal ternary phase diagram of a crosslinked polymer (2), solvent (1), and soluble oligomer (3). Coexisting concentrations $A_1 B_1$, $A_2 B_2$ etc., points A_1, A_2 etc. are on the swelling curve; tie lines $A_2 B_2$ etc. passing as prolongations through point C are quasibinary sections; arrows indicate the change in concentration in the ultracentrifugal field; $w_i^\alpha =$ the mass fractions of i in phase α (see text)

3.3. Ternary state diagram

The situation of the heterogeneous swelling equilibrium is illustrated using the triangular coordinates of W. Gibbs. The ternary system is represented schematically in Fig. 2, where the corners of the equilateral triangle correspond to the pure components. At the lateral sides the three possible binary systems are described by the mass fractions of i, where the mass fraction w_i^α in phase α is given by $w_i^\alpha = \rho_i^\alpha / \rho^\alpha$.

We assume the crosslinked polymer to undergo a swelling equilibrium with the pure solvent. The points of coexistence are A_1 and B_1. In adding the oligomer to the binary system different ternary swelling equilibria represented by the points A_k, B_k are possible which are connected by the tie lines. The geometrical locus of all swelling concentrations due to the points A_k is the swelling curve $A_1, A_2 \ldots A_7$. At A_7 the pure oligomer is the swelling agent of the crosslinked polymer. For simplicity it is assumed that the prolongations of all tie lines pass through the corner C. This means that all tie lines are quasi binary sections, along which the ratio of solvent and oligomer content is constant. If there is a different ratio of solvent and

oligomer in the gel there is a different slope of the tie lines with respect to the quasibinary sections.

If at the beginning of the sedimentation experiment a gel and solution are present due to the concentrations in points A_2 and B_2, we know that first the gel has to build up a concentration gradient, if the centrifugal field is acting. As A_2 corresponds already to the swelling equilibrium the concentration of the polymer can only be increased in direction of C leading to A_2' at fixed equilibrium concentration belonging to A_2 at the gel surface. Therefore, the mass conservation will force the gel to move the phase boundary gel/solution to the bottom of the cell as mentioned. The increase of the concentration of component 2, w_2'', at the bottom of the cell is indicated with an arrow indicating point C. During the formation of gradients of the concentrations in the gel there is also the beginning of the set up of the gradients of the concentrations in the solution, which is also indicated by an arrow around B_2 reaching finally from B_2' to B_3. As the gel surface is in contact with the highest concentration of the oligomer, we expect a coexistence between points B_3 and A_3. Therefore, the concentration of the gel at the boundary gel/solution has to change from A_2 to A_3 in the course of attaining the final equilibrium state. But inside the gel the concentrations are between A_3 and A_3' due to the action of the external field. The corresponding mass fractions belonging to A_3' are represented as distances at the axis of the binary system of components 1 and 2. As both continuous equilibria and also the heterogeneous between B_3 and A_3 have to establish, one has to take the small polymer layer in order to achieve equilibria in relatively short times.

3.4. Mass conservation

If the gel extends between cell bottom (r_2) and phase boundary vapor/solution (r_1) at $t = 0$, we may define a function of $\rho_2(r)$ representing an area in a diagram, which has to be the same for all movements of the gel surface.
This function is

$$\frac{1}{r_2 - r_1} \int_{r_1}^{r_2} \rho_2 r \, dr = \rho_2^0 \frac{r_1 + r_2}{2}, \qquad (13)$$

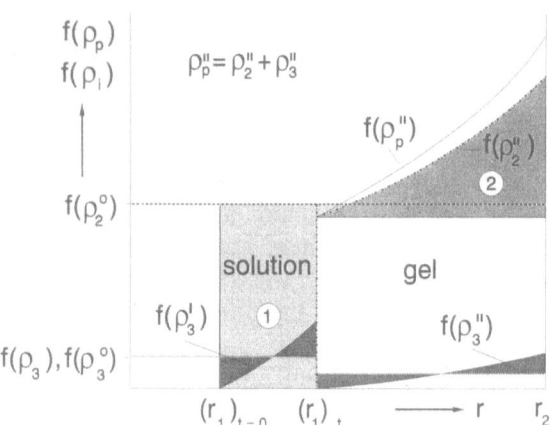

Fig. 3. Schematic isothermal concentration/distance diagram of the oligomer or soluble polymer (3) and the crosslinked polymer (2); ρ_3^α, ρ_2^α = partial densities of oligomer and crosslinked polymer in phase α; $f(\rho_i)$ = concentration function defined by mass conservation (see text)

which is plotted as $f(\rho_i)$ vs r in Fig. 3. The original concentrations are indicated by ρ_i^0. Thus, the concentration of component 2 is given at time $t = 0$ by the righthand side of Eq. (13) multiplied by $(r_2 - r_1)$ which gives $\rho_2^0 (r_2^2 - r_1^2)/2$. In Fig. 3 the value $f(\rho_2^0)$ corresponding to ρ_2^0 is represented by a dashed line. If the concentration gradients in the gel have established and the boundary of the gel/solution has moved to $(r_1)_t$ the dashed areas (1) and (2) have to be the same. The integral of component 2 due to Eq. (13) is represented by the dashed dotted curve.

Additionally, also a much smaller concentration of the oligomer has been plotted by dotted lines for time $t = 0$ in both phases. The corresponding values at time t are demonstrated by thin full curves. The addition of components 2 and 3 results in the local dependence of the total amount of the oligomer + crosslinked polymer, $\rho_p = \rho_2 + \rho_3$, which is schematically shown by the strong full curve in the upper part of the figure belonging to the gel phase.

The quantitative description of mass conservation starts with the presumption that the total mass of every component i has to stay in the ultracentrifugal cell. From the original concentration of the system at time $t = 0$, named ρ_i^0, the concentration of i at every distance from the axis of rotation, $\rho_i(r)$, may be calculated using interference or Schlieren optical systems for the detection of the refractive index or the gradient of it [5]. For

a better reading of the distances, we now introduce $r_1^0 \equiv r_m^{g/v}$ and $r_1 \equiv r_m^{g/s}$, where r_2 remains the value of r for the bottom of the cell.

Mass conservation of the swollen crosslinked polymer in a sector-shaped cell is described by

$$\phi h \int_{r_1^0}^{r_2} \rho_2^0 r \, dr = \phi h \int_{r_1}^{r_2} \rho_2''(r) r \, dr . \tag{14}$$

Dividing by the sector angle ϕ and the height of the cell h and integrating the lefthand side of Eq. (14) results in

$$\int_{r_1}^{r_2} \rho_2''(r) r \, dr = \rho_2^0 \frac{r_2^2 - (r_1^0)^2}{2} . \tag{14a}$$

For the soluble polymer the corresponding integral of the concentration gives for the beginning of the experiment

$$\int_{r_1^0}^{r_2} \rho_3^0 r \, dr = \rho_3^0 \frac{r_2^2 - (r_1^0)^2}{2} . \tag{15}$$

If the boundary gel/solution has formed the mass conservation reads

$$\int_{r_1^0}^{r_2} \rho_3^0 r \, dr = \int_{r_1^0}^{r_1} \rho_3'(r) r \, dr + \int_{r_1}^{r_2} \rho_3''(r) r \, dr . \tag{15a}$$

The first integral of the righthand side of Eq. (15a) can be obtained directly by an interference optical system. As the lefthand side of Eq. (15a) may be determined by dialysis for thermoreversible gels at an appropriate choice of temperature [11], the second integral on the righthand side of Eq. (15a) may be solved numerically assuming special dependence of the concentration gradient of component 3 on distance r or by direct measurements, perhaps using UV-absorption optics and marked oligomers showing UV-absorption in combination with Schlieren or Rayleigh optics.

If the concentrations have been determined by means of a Schlieren optical system, the integral of the Schlieren-distance curve yields for the swollen crosslinked polymer [5]:

$$\int_{r_1}^{r} \frac{d\rho_2''}{dr} \, dr = \rho_2''(r) - \rho_2''(r_1) \quad \text{for } r_1 < r < r_2 . \tag{16}$$

Now, taking into account Eqs. (14) and (14a), we deduce from Eq. (16) by multiplication by r and integration in the notation of Fujita:

$$\rho_2''(r_1) = \rho_2^0 - \frac{2}{r_2^2 - r_1^2} \int_{r_1}^{r_2} r \left[\int_{r_1}^{r} \left(\frac{d\rho_2''}{d\zeta} \right) d\zeta \right] dr \,,$$

(17)

where ζ is the distance variable instead of r. If $\rho_2''(r_1)$ is known, $\rho_2''(r)$ can be calculated for every distance by use of Eq. (16).

In the same way, we get for component 3:

$$\int_{r_1^0}^{r} \frac{d\rho_3'(r)}{dr} \, dr = \rho_3'(r) - \rho_3'(r_1^0) \quad \text{for } r_1^0 < r < r_1$$

(18a)

and

$$\int_{r_1}^{r} \frac{d\rho_3''(r)}{dr} \, dr = \rho_3''(r) - \rho_3''(r_1) \quad \text{for } r_1 < r < r_2 \,.$$

(18b)

Multiplying by r, integration and use of Eq. (15a) leads to

$$\rho_3'(r_1^0) = \rho_3^0 - \rho_3''(r_1) \frac{r_2^2 - r_1^2}{r_1^2 - (r_1^0)^2}$$

$$- \frac{2}{r_1^2 - (r_1^0)^2} \left\{ \int_{r_1^0}^{r_1} r \left[\int_{r_1^0}^{r} \frac{d\rho_3'}{d\zeta} \, d\zeta \right] dr \right.$$

$$+ \left. \int_{r_1}^{r_2} r \left[\int_{r_1}^{r} \frac{d\rho_3''}{d\zeta} \, d\zeta \right] dr \right\} \,.$$

(19)

From Eq. (19) it can be seen that $\rho_3'(r_1^0)$ and $\rho_3''(r_1)$ are coupled. Therefore, it is advantageous to determine $\rho_3^\alpha(r)$ with $\alpha = '$ and $''$ by a second optical system separately.

A special case may be that only the sum of the concentration gradients of components 2 and 3 can be detected by the Schlieren optical system.

Then, we may have $\rho_p'' = \rho_2'' + \rho_3''$ and, consequently,

$$\frac{d\rho_p''}{dr} = \frac{d\rho_2''}{dr} + \frac{d\rho_3''}{dr} \,.$$

(20)

First, $\rho_3''(r)$ has to be determined. Afterwards the gradient $d\rho_2''/dr$ can be calculated from Eq. (20) and resolved by use of Eq. (17).

Acknowledgements

The author thanks the Deutsche Forschungsgemeinschaft for financial support of the project and Dipl.-Chem. H. Cölfen for discussion and drawing the figures.

References

1. Borchard W (1991) Progr Colloid Polym Sci 86:84
2. Borchard W, Cölfen H (1992) Makromol Chem Makromol Symp 61:143
3. Svedberg T, Pedersen KO (1940) Die Ultrazentrifuge. Steinkopff-Verlag, Dresden
4. Haase R (1963) Thermodynamik der irreversiblen Prozesse. Steinkopff-Verlag, Dresden
5. Fujita H (1962) Mathematical Theory of Sedimentation Analysis
6. Haase R (1956) Thermodynamik der Mischphasen. Springer-Verlag, Berlin-Göttingen-Heidelberg
7. Rehage G, Meys H (1958) J Polym Sci 30:271
8. Scholte Th G (1968) J Polym Sci (A-2) 6:91 (1970) European Polym J 6:51
9. Boyer RF, Spencer RS (1948) J Polym Sci 3:97
10. Tompa H (1956) Polymer Solutions Butterworths Scientific Publication, London
11. Cölfen H (1993) dissertation, Duisburg

Received June 18, 1993;
accepted December 6, 1993

Author's address:

Prof. Dr. W. Borchard
Universität GH-Duisburg
Angewandte Physikalische Chemie
Lotharstraße 1
47048 Duisburg

Progress in Colloid & Polymer Science   Progr Colloid Polym Sci 94:90–101 (1994)

A modified experimental set-up for sedimentation equilibrium experiments with gels. Part I: The instrumentation

H. Cölfen and W. Borchard

Angewandte Physikalische Chemie der Universität-GH-Duisburg, Duisburg, FRG

Abstract: An experimental set-up for the determination of swelling pressure-concentration curves and thermodynamic and elastic properties of gels by means of analytical ultracentrifugation is presented. The instrumentation described allows an accurate measurement of up to 70 samples simultaneously in an equilibrium experiment which lasts about 2 days if the 10-hole multichannel centerpieces described in part 2 of this trilogy are used. The number of samples is enough to characterize one complete gel/solvent system by important thermodynamic, structural and elastic parameters in the experimentally accessible range. It is of great advantage that all samples are measured under exactly the same conditions. This is not possible by means of classical methods which all yield only one swelling pressure corresponding to one concentration. The evaluation process of the Schlieren patterns is fully automated to ensure a user-friendly process.

Key words: Gels – analytical ultracentrifugation – sedimentation equilibrium – swelling pressure – computerized Schlieren pattern evaluation

Introduction

The increasing importance of gels in scientific research and industrial applications recommends the determination of their physical and thermodynamic properties [1]. The investigation of both chemically and physically crosslinked gels can be done very advantageously by means of equilibrium analytical ultracentrifugation because it is possible to derive the dependence of the swelling pressure or the difference of the chemical potential of the solvent between gel and the pure solvent on the polymer concentration in a large range of concentrations. Furthermore, thermodynamic and elastic properties of gels can be derived [2–7]. If a multiplace rotor and/or multichannel centerpieces are used, it is possible to obtain several swelling pressure-concentration curves which are measured under exactly the same conditions in one equilibrium experiment. Such an amount of valuable information on gels cannot be derived by means of any of the classical methods because they have the great disadvantage to yield only

a single swelling pressure corresponding to one concentration. Brief reviews of classical devices for the determination of swelling pressures of gels with their limitations and the history of ultracentrifugal investigations on gels are given in [3, 7].

Up to now, the experimental conditions allowed only the measurement of as much gel samples as rotor places were available in one experiment. But even when using a six-hole rotor, which is the analytical rotor with the largest number of holes available for the model E centrifuge, the measuring procedure is really ineffective due to the very long duration of up to 3 weeks for one equilibrium experiment with gels [2–10]. The high column lengths of up to 1 cm, which led to the long experimental duration, had to be used to minimize errors due to the very limited graphic capabilities of a picture evaluation with an Apple IIe computer. So, several solutions have to be analyzed to increase the efficiency of the centrifugal experiments and to make the method quick and powerful. Some requirements were already realized in recent decades of ultracentrifugation such as the development of multichannel cells and the short column technique [11], which take advantage of the finding of van Holde and Baldwin that the time which is needed to reach the sedimentation equilibrium is proportional to the square of the filling height of the cell [12]. The short column technique has been developed to a rapid and sure method to investigate macromolecules in solution [13].

This paper is the first of three dealing with significant improvements in the efficiency of the measuring procedure for the example of gels. The following papers will present technical developments and experimental results [14, 15]. Every scientist who investigates solutions should derive some ideas therefrom to apply parts of this advantageous procedure to experiments with solutions.

The described system was developed as inexpensively as possible. Major developments are a 10-hole short column cell optimized for the Schlieren examination of gels, an eight-hole rotor for the model E, a programmable multiplexer, an improved modulation system for a continuous laser, a modified Schlieren optical system for 10-hole centerpieces, the installation of a fully automatic picture digitization system, and a software package with mainly fully automatic evaluation programs.

The components are tested with the Beckman model E ultracentrifuge. It should be no general problem to apply it to the MOM, MSE CENTRISCAN or the MSE MK II ultracentrifuges. In the new Beckman Optima XL-A the system is predicted not to work because of the present lack of Schlieren optics, although the application of a programmable multiplexer would enable the use of different analytical rotors in this centrifuge beside the present four-place rotor. When presenting the developments, the authors try to critically mention the pros and cons of the introduced system. Therefore, the results of the modifications are discussed immediately at their presentation to ensure an easy reading of this article.

Short description of equilibrium experiments with gels

The sedimentation equilibria of gels in an ultracentrifugal field are described or treated in detail elsewhere [2–10, 16–21]. Therefore, only a short description of the method is given to ensure that the reader is able to understand the principle. If gels are investigated by means of equilibrium runs in an Analytical Ultracentrifuge, the centrifuge is used as an osmotic pressure generator. The radial pressure gradient causes a corresponding concentration gradient in the gel phase. As soon as the maximum solvent concentration due to the swelling equilibrium is established at the meniscus gel/vapor, a sol phase is formed and the new meniscus gel/sol begins to move towards the cell bottom until an equilibrium position is reached. From this equilibrium position it is possible to calculate the dependence of the swelling pressure on the polymer concentration, which is called swelling pressure-concentration curve in a large concentration range [2–10, 16]. The changes in the polymer concentration and the movement of the meniscus gel/sol are observed with the Schlieren optical system. As supplementary measurements density measurements of the gels have to be made to determine the dependence of the gel density on the polymer concentration [22]. Furthermore, swelling measurements are necessary to obtain the polymer concentration of the maximum swollen gel.

Instrumentation

a) The modulation system

Generally, the light source or the detection system of an Analytical Ultracentrifuge has to be modulated if a multihole rotor (four, six or eight-hole rotor) is used. In past years several different techniques were suggested. The virtue of a modulated light source is discussed in [23].

The modulation system used in this work is shown in Fig. 1. It is partly based upon the system installed by Holtus on the model E centrifuge of the Duisburg work group [8].

The rotor code ring was painted black in such a way that only one part is left metallic for reflection. The ring is illuminated by an IR-LED. The diode is placed together with a phototransistor in a small housing with focusing optics (MRL 601, Conrad Elektronik, Hirschau, FRG) under the rotor support fork. Every time the metallic part of the code ring passes by, a reflection is caused. It is

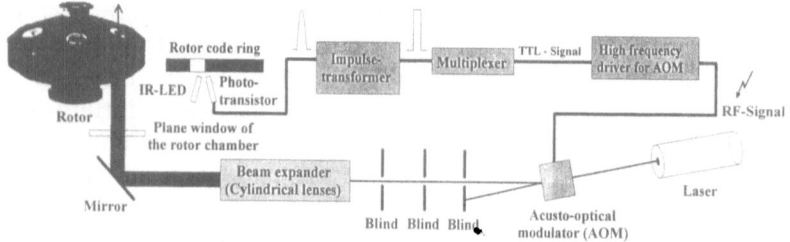

Fig. 1. Modulation system applied to the use of a multihole rotor illuminated by laser light.

detected by the photo transistor which generates an electrical impulse. The impulse is transferred to an impulse transformer which modifies the incoming signal to a rectangularly shaped impulse [14]. Therefore, every rotation one impulse is transferred to the multiplexer which generates a TTL – signal with a defined time delay to the incoming signal [14]. The TTL signal is the incoming signal for the high-frequency driver (Newport N21080-1.5 DS) of the acusto-optical modulator (AOM). The driver sends a RF-signal (RF = radio frequency) with a frequency of 80 MHz and a power of about 1 W to the AOM (NEOS N23080-2). If this signal is sent, the beam of laser light passing the modulator is diffracted under a certain angle called the Bragg angle.

Only the diffracted light is able to pass a set of three blinds which ensures that each measuring cell is illuminated by the laser light exactly when it passes the light path. Normally, the same effect could be achieved by using only a single blind, but the three blinds are useful to align the laser light beam exactly parallel and in the middle of the optical bench. This ensures that the light beam passes all further lenses which are placed on the same bench exactly in the middle if each lens is aligned properly. The parallel circular cross-section of the round light beam with a diameter of 2 mm is expanded by factor 10 in one direction by a beam expander consisting of two cylindrical lenses (f = 10 mm, f = 100 mm, Spindler & Hoyer, FRG) to a small band of parallel light. The optics is described in detail in part 2 of this trilogy [14]. The light is deviated by a mirror at an angle of 90°. It is important that the cylindrical lenses are sized large enough that the light does not pass the side regions of the lenses. In the case of too small lenses, both sides of the light band should be cut off by means of a blind. Otherwise, intense optical errors are caused which result in a disturbed baseline in the Schlieren patterns. After passing the plane window of the rotor cham-

ber, the light band illuminates the selected measuring cell if this cell reaches the position of the light beam.

It is advantageous to apply an AOM as modulator because of the very fast modulation in the magnitude of nanoseconds which is possible compared with pulsed lasers or stroboscope flashlights. The fast modulation ensures a high resolution in the choice of locations of the light flashes on the rotor because the rotor changes its rotational position much less than 1° even at the highest rotational speed of 68 000 revolutions per minute (RPM) until the next light flash can be set.

But there are several disadvantages in applying an AOM which all result in limited light intensities of the pictures on the screen. At first an AOM allows a limited power of laser light. The originally applied modulator, the Newport N24080, allows 30 mW/mm^2 only. Therefore, the power of the laser which can be applied has an upper limit. The active height which limits the diameter of the laser light beam is 1 mm in this AOM. Therefore, light beams with larger diameter have to be focused to gain maximum laser power. But this procedure very much endangers an AOM with a low limited laser power density if the diameter of the laser beam is not measured exactly. In such a case the maximum power density may be exceeded. On the other hand, most of the laser power cannot be used without this necessary step.

A further disadvantage is the light absorption and the rather low deflection efficiency of the modulator. The maximum deflection efficiency of a modulator depends on the diameter of the light beam. For the Newport N24080 AOM it is 65% which means only 65% of the incoming light can be used as pulses of light deviated with the Bragg angle even if the modulator is properly aligned to the first order of deflection. A general disadvantage of the modulation technique is that only a very small part of the light can be used for the illumination of the cells. If one revolution of the rotor is divided into 1000 steps by the multiplexer, ensuring an angle resolution of better than 0.3°, only 0.1% of the pulsed laser power can be used. Even if 14 sequential flashes illuminate nearly the maximally possible segment in an analytical cell of 3.9°, the amount of light used to illuminate the cell is only increased to 1.4%. Combined with the deflection efficiency of 65%, this means that, at maximum, 0.91% of the original power of the laser can be used to illuminate the cell. Furthermore, the large number of lenses, at least in the Schlieren or interference optics, causes further light absorption. So, one can imagine that the described modulation system works hard at its limits.

The application of the high power AOM (NEOS 23080-2) with a maximum power density of 3 W/mm², an active height of 2 mm, and a maximum deflection efficiency of 92% in combination with a 40 mW low-cost He–Ne laser (HNC 4000 s, ES-Lasersysteme, Mössingen, FRG) ensures a good evaluation of Schlieren patterns by means of combined video and computer techniques. The application of modern picture digitizing techniques requires a much higher light intensity than is needed for the much more sensitive photographic technique because the latter works with exposure times.

Modern high power laser diodes (50–100 mW) with $\lambda \leq 670$ nm, which are now available, seem to be the best light source for the investigation of gels with Schlieren optics. They are fast modulable, quite cheap, small in size, and have a very long lifetime. Disturbing interferences in the Schlieren patterns, as known from solutions, have never been observed in the patterns of gels although a He–Ne laser with a larger coherence length than that of a laser diode and especially that of a mercury lamp has been used as light source.

b) The eight-hole titanium rotor

An analytical eight-hole titanium rotor of a Heraeus AZ 9100 centrifuge was modified for the use in the Beckman model E. This rotor had a maximum rotor speed of 72 000 RPM before the modification from the standing to the hanging operation. The modification was mainly achieved by introducing a new rotor axle in the center hole

of the original Heraeus rotor with the suspension to the model E drive. As the rotor was too high to enable an operation in the model E, both the top and the bottom of this rotor had to be shortened so that the rotor has the height of an original Beckman rotor for the model E. In order to not change the counterbalance of the rotor too much, the height of material which had to be taken away from the top and the bottom of the rotor had to be calculated in a way that the masses of the removed titanium at both ends are the same. Furthermore, stability calculations had to be performed to guarantee that the modifications of the rotor ensure a proper operation up to the maximum speed of 40 000 RPM needed for the equilibrium experiments with gels.

The new axle with the diameter of the screw ring and its thread at the top and with a larger platform at the bottom to hold the rotor was manufactured of high quality stainless steel. A schematic diagram of the rotor is presented in Fig. 2.

Calculations concerning the expansion coefficients of steel and titanium ensured the use of the rotor in the experimentally interesting temperature range for the investigation of gels or most of the solutions from 0° to 40 °C. The use of titanium as material for the center axle of the rotor was not

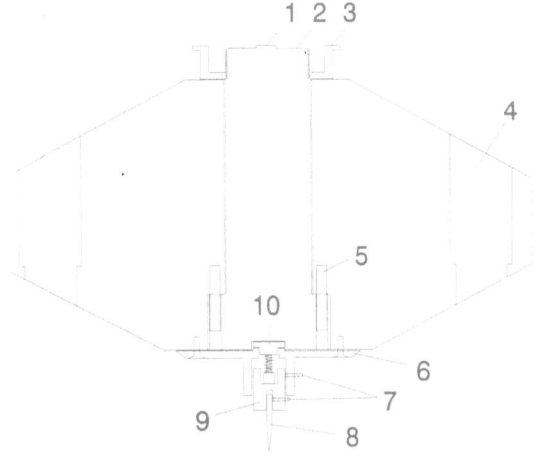

Fig. 2. Schematic diagram of the modified eight-hole titanium rotor with 1 = coupling stem, 2 = rotor axle with thread for the driving axle, 3 = support ring (On this ring the rotor code ring is placed), 4 = cell hole, 5 = tightening pins for the rotor axle, 6 = thermistor mount, 7 = fastening screws, 8 = thermistor needle, 9 = thermistor holder and 10 = thermistor.

necessary because the centrifugal force acting in the center range of the rotor is relatively low compared with that acting at the unmodified periphery. Calculations for the maximum speed were made to ensure a safe operation of the modified rotor in the model E centrifuge, especially under the aspect that no manufacturer was able to give a guarantee on the maximum rotational speed as well as on the lifetime etc. of the rotor. Due to extreme caution it was decided to design the rotor with a nearly threefold safety range. The rotor was counterbalanced at 2000 RPM with an accuracy of 3.46 g mm. This value corresponds to a mass difference of only 40 mg at the periphery of the rotor. This is much more than necessary if one keeps in mind that the differences in the masses of the measuring cells are allowed to reach 500 mg.

c) *The digitization hardware*

The evaluation of Schlieren pictures from the photographic system of the ultracentrifuge is time-consuming and ineffective. But only the rapid development of powerful computer techniques in the 1980s enabled first possibilities for centrifuge users to replace the photographic detection system of the Schlieren optics by half- or fully automatic picture digitizing systems which basically consist of a screen, a video camera, a frame grabber, and a computer [7, 8, 10, 24, 25].

Due to the lack of sufficient software, the programming of the whole software or at least the modification of existing digitization software had to be done by the users themselves. Also, the digitizing boards, the so-called frame grabbers, as the digitization software were very expensive because they were only used in very special applications. Only the introduction of graphical user surfaces for the widespread PC's like Microsoft-Windows, whose current version 3.1 is even

capable of multimedia features, opened the field of video digitization for a large number of users. The use of frame grabbers for multimedia or desktop publishing applications, just to name a few, made the frame grabbers inexpensive due to the large market and competition in that field. Today, many sufficient frame grabbers including digitization software are available at low prices so that every user should find his product of choice. According to the rapid developments in computer and software techniques it should be possible to write on-line evaluation programs for different kinds of ultracentrifugal experiments as Laue did for interference patterns [26, 27] and Mächtle partly realized for Schlieren patterns [25].

Fortunately, a so-called on-line technique is not required by efficient equilibrium experiments with gels because only the Schlieren patterns at the very beginning of the experiment, such as those in the equilibrium case, are necessary for a proper evaluation. All other pictures are just useful for the documentation of effects of any kind during the experiment or to see whether the equilibrium is reached already or not.

The picture digitization system used with both model E ultracentrifuges in Duisburg works fully automatically in contrast to the half automatic one which was used before [2–10]. It is shown schematically in Fig. 3.

The fully automatic digitization has the advantage to be more comfortable, objective and capable of on-line data analysis compared to half automatic systems which digitize pictures, for example, by means of a graphic tablet. A further advantage is that all patterns may be stored for documentation as a picture similar to a photograph. A disadvantage of the fully automatic system is that a higher light intensity on the screen is needed for a proper digitization of the pictures. If the light intensity is subcritical, the screen may be

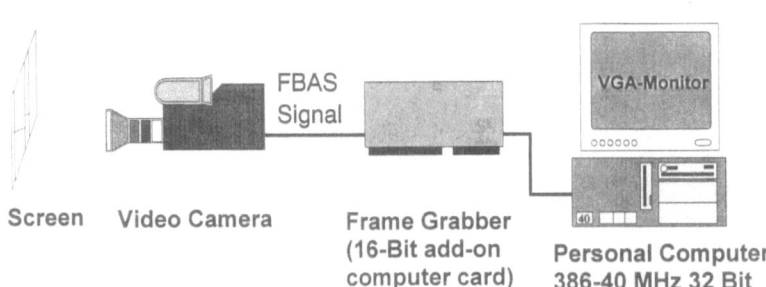

Fig. 3. Fully automatic picture digitization system for Schlieren or interference patterns based upon a video camera which is connected to a personal computer.

removed and the image can be detected directly by the camera optics.

After the Schlieren or interference patterns are imaged on a screen which is placed in the film plane of the former photographic system, the picture is filmed with a light sensitive (0.1 lux) monochrome vidicon video camera (Ikegami model ITC-410, Ikegami Tokyo, Japan) with a resolution of 650 (horizontal) * 625 (vertical) pixel. Although the geometric stability of vidicon cameras is less than that of CCD-cameras (CCD-cameras have light-sensitive elements of well defined size, shape and spacing) the vidicon camera has been chosen due to the higher light sensitivity. The camera transfers a standard FBAS video signal to the frame grabber (Micro Eye 1c Digithurst, Nürnberg, Germany) which is placed in the computer. This frame grabber ensures the fully automatic picture digitization with the program Autoscope (Digithurst, Nürnberg, Germany) with a resolution of 640 * 480 pixel in 256 colors or gray scales. The program further allows some picture quality improvement like contrast enhancement or brightness control. Further picture evaluation is done with a commercial PC-AT 386-40 MHz with 4 Mbyte RAM (Random Access Memory). This computer is equipped with a super VGA monitor capable of a resolution of 1024 * 768 pixel (AOC) and the fast Diamond Speedstar super VGA card with the same resolution as the monitor.

Figure 4 shows a Schlieren pattern of an experiment with a solution in a monosector cell immediately after the digitization. The picture is digitized in the standard VGA resolution of 640 * 480 pixel with 256 gray scales. The markings of the reference holes of the counterbalance cell (5.7 and 7.3 cm) are located on the very left- and righthand sides of the 640 pixel axis. As can be seen, this number of gray scales ensures a good picture quality. The resolution of 640 * 480 pixels has been chosen due to the file size of the latter digitized pictures and the resolution of the video camera of 650 * 625 pixel. As in every picture digitization process, the user has to make a compromise between the resolution of the picture and the amount of disk space which is needed to store the digitized picture. As the number of pictures which have to be stored is very large due to the efficiency of the measuring procedure, the file size of the pictures is of great importance. Another

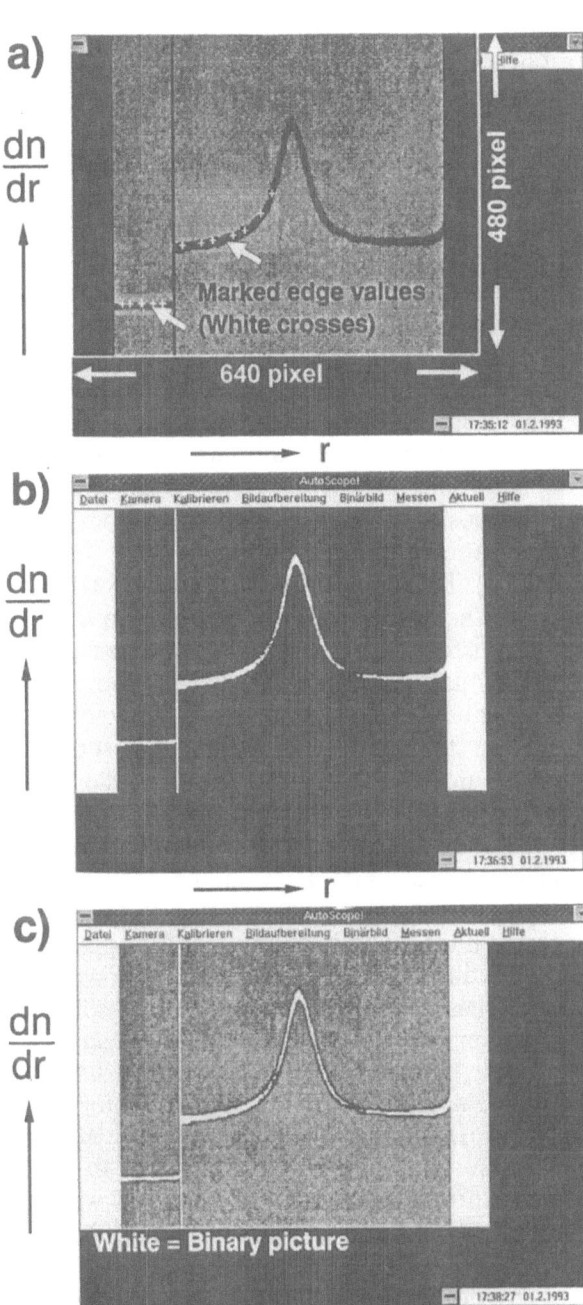

Fig. 4. a) Screenshot of a digitized Schlieren pattern of sedimenting polyvinyl pyrrolidine in H_2O (c = 4 g/l, 40 000 RPM) in a monosector cell in the standard VGA resolution of 640 * 480 pixel in 256 gray scales (8-bit). The radial range of 1.6 cm between the reference marks of the counterbalance cell is covered in the horizontal. Also, edge values of a range of gray scales for the conversion to a 1-bit black & white picture are marked as white crosses. n is the refractive index and r the distance from the axis of rotation b) Screenshot of a binary Schlieren pattern 1-bit black & white converted from the picture a). c) Screenshot of combined 256 gray scale and binary picture.

reason for the application of the standard VGA resolution is that further picture evaluation programs were partly written under Microsoft Quick Basic 4.5. This programming language just allows, at maximum, a screen resolution of $640 * 480$ pixels with 16 colors.

The digitizing resolution of $640 * 480$ pixels for a distance of 1.6 cm between the two reference holes of the counterbalance cell results in a horizontal resolution of $25 \mu m$/pixel. If the camera displays more than the 1.6 cm between both reference marks of the counterbalance cell, the resolution is a little bit poorer. But this resolution should be enough for an accurate evaluation of equilibrium experiments with gels, because up to now it is only possible to evaluate the positions of the phase boundaries as the concentration gradient is not visible in the whole gel phase [7]. Therefore, the vertical resolution is irrelevant to the results. An error of 1 pixel in the horizontal direction for a typical experiment with a 8.73% gelatin gel (14 000 RPM, 0.8 cm filling height) causes a maximum relative error of about 0.03% in the calculated swelling pressures and a maximum relative error of 0.4% in the corresponding polymer concentration. An error of 3 pixels results in maximum relative errors of 0.03% in the swelling pressure and 1.2% in the corresponding polymer concentration. The latter errors can be expected if the 10-hole centerpieces are applied, because in this case the filling height is lower than 0.3 cm. These low errors due to the resolution of the digitization system enable the use of the short column cells. Considering the mentioned errors, it should be noted that even the results (filling height up to 1 cm) which have been achieved with an Apple IIe computer with about one-third of the video camera's horizontal resolution agree very well with literature data which have been derived by other methods [4, 7].

It should be no major problem to apply higher resolutions, because under Microsoft Windows 3.1 some powerful programming languages like Microsoft Visual Basic Professional 3.0 are available which should enable an easier, more powerful and comfortable programming of picture evaluation programs.

d) The picture evaluation process

It is of great advantage if the 8-bit 256 gray scale pictures are converted to a 1-bit black & white picture. This can be done by marking edge values for the range of gray scales which have to be converted to white. The marked edge values are shown as small white crosses in Fig. 4a.

The conversion follows by means of a simple mouse click. The converted binary picture is presented in Fig. 4b. It is obvious that this binary picture is much easier to evaluate than the 256 gray scale picture because the information in this picture is reduced to that which is necessary for the further evaluation. Furthermore, the amount of disk space for the file storage is reduced significantly.

It is possible to control whether the picture conversion was successful or not by combining the binary and the gray scale picture. This combination is shown in Fig. 4c. The white areas belong to the binary picture. As it can be seen, the conversion was really successful. Otherwise, the conversion can be repeated by selecting new edge values. This may be wished if the Schlieren curve in Fig. 4b should be continuous. The binary picture is stored as BMP (windows bitmap) or TIFF (tagged image format) file. Up to this point, the described procedure works for gels and solutions. The evaluation software described in the following is especially written to evaluate equilibrium experiments with gels. There are two versions of the basic programs because the picture evaluation differs for sector-shaped cells or multichannel cells, which will be presented in part II [14].

The number of pictures which are stored during an equilibrium experiment is evaluated by means of a software package presented in Fig. 5. At first, the basic program Multicon is applied which extracts experimental parameters like the system, the picture number, the cell number and the concentration of the sample from the file names. It is clear that this information has to be stored as a code because the length of the filename under DOS is restricted to just eight digits due to the limitations of the MS-DOS operating system.

After this extraction, Multicon requests the single input of further experimental data like the start time of the run, the time of its end, and the corresponding readings of the mechanical counter for revolutions, etc. From these data the program calculates the average rotational speed, the elapsed time from the beginning of the experiment to the moment the Schlieren picture is digitized and other data. All experimental data of the selected

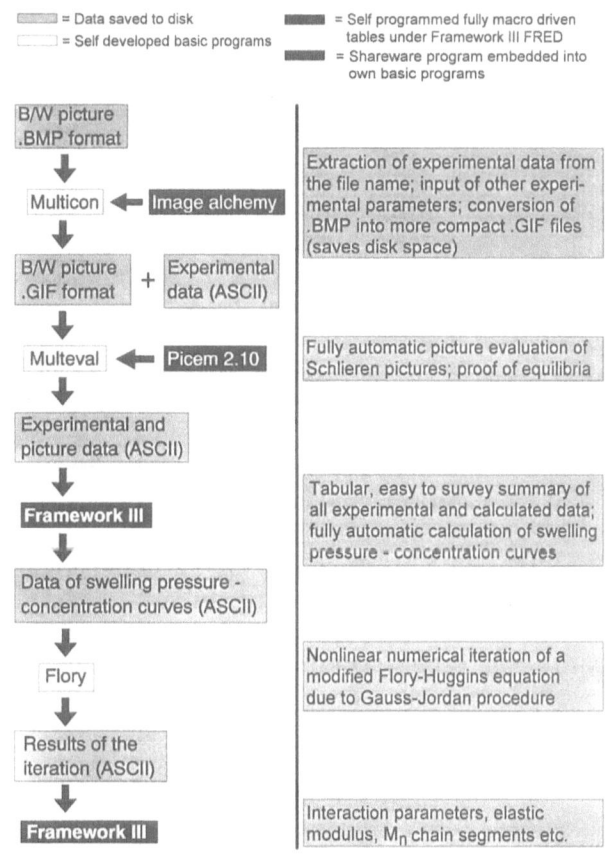

Fig. 5. Schematic representation of the software package for the fully automatic evaluation of Schlieren pictures from equilibrium experiments with gels in 10-hole multichannel cells. An analogue software package has been written for sector shaped cells.

pictures are stored in an ASCII file. In order to save disk space, the Schlieren patterns are converted from the BMP into the more compact GIF (Compuserve) format by means of the shareware program Image Alchemy 1.5 (Handmade Software, Inc.) which is embedded into the Multicon program.

Now, the program Multeval is loaded and the Schlieren patterns are evaluated fully automatically. The program is able to evaluate series of pictures as well as single pictures. After a Schlieren pattern is loaded by means of the embedded shareware program Picem 2.1, Multeval determines the color value of each pixel in 30 horizontal lines as is shown in Fig. 6. Every time a change from black to white or reverse is detected, Multeval stores the horizontal position in the RAM of the computer.

After all color values of the pixels in the 30 horizontal lines are determined, the program compares the values of the black/white boundaries. The position of the edge of a meniscus is obtained by means of a threshold value which has to be exceeded by the number of equal horizontal positions of the black/white boundaries for that meniscus. If, for example, the threshold value is 10, at least 10 positions for a black/white boundary of the 30 horizontal lines of pixels have to be equal. This procedure is due to the fact that only a meniscus has a vertical boundary. Such a procedure requires that the video camera is aligned properly so that a vertical meniscus is displayed

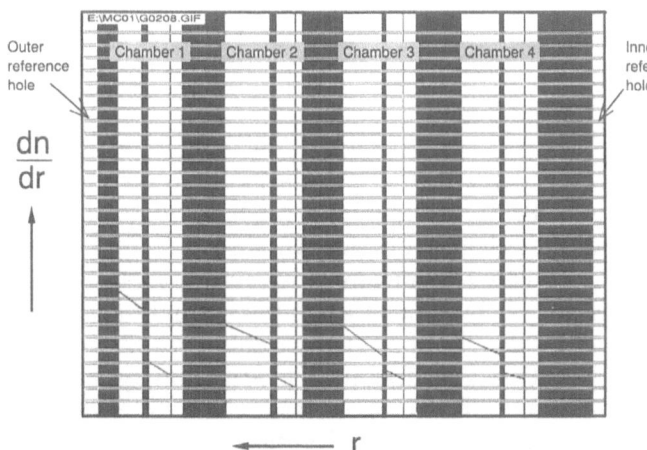

Fig. 6. Fully automatic evaluation of Schlieren patterns of gel samples in multichannel cells with the program Multeval. A test picture for a row of four sample chambers is evaluated. The horizontal lines mark both sides of the 30 lines for which the color values of the pixels are determined. The filename in the upper lefthand corner consists of the path name and a coded name of the image file with the extension GIF. n is the refractive index and r the distance to the axis of rotation.

vertically in the digitized picture. After the boundaries of the menisci are determined, they are marked. Now the positions of the idealized menisci are calculated due to considerations presented in [14]. These positions are stored to disk with all other relevant experimental parameters and the data from the picture evaluation in an ASCII file.

If the Schlieren pictures are digitized rather poorly, for example, if the light intensity of the picture is low due to the turbidity of the sample, the boundaries of the menisci are not so sharp as in the test picture in Fig. 6. In these cases the fully automatic evaluation process does not work. Therefore, Multeval has a routine to evaluate the pictures by hand. The boundaries of the menisci are marked with a graphic cursor which is controlled with the cursor buttons of the computer keyboard.

After all Schlieren patterns of one experiment are evaluated, Multeval determines the shift of the meniscus gel/sol to the equilibrium position. By means of a relative displacement, which was introduced by Holtus [8], the equilibria can be proved with the principle of reaching an equilibrium in several ways. The equilibrium position of the meniscus gel/solvent has to be reached when the rotational speed is increased to the selected value as also when the rotational speed is decreased from a higher value to the selected one after an obvious shift of the meniscus gel/solvent towards the cell bottom has taken place. Alternatively, the more general proof of equilibria by means of coinciding swelling pressure-concentration curves can be applied [7].

Multeval has routines which calculate the swelling pressure–concentration curves as all other parameters which can be derived from these curves. If these calculations are done, the user is able to look at the experimental results in an early stage of the evaluation procedure. In the authors' experience, not all swelling pressure–concentration curves are determined correctly due to different reasons. The user is able to detect these wrong results so that they have not to be read into the Framework III tables. This helps to reduce the immense amount of data.

After all pictures have been evaluated, the data from the picture evaluation as the corresponding experimental data are read into a Framework III table. Framework III (Ashton Tate, now Borland) is a table calculation program which ensures

a tabular summary of all experimental and calculated data which is easy to survey. A great advantage of the application of a table calculation is the structure of the calculation tables. If a formula for the calculations should be modified, only the formula itself has to be changed. All other calculations with the new formula are done just by the press of one button. Furthermore, such a table is an open structure which can be modified or extended without changing the data set.

As with many other table calculations, Framework has a specific editor language, which is called FRED (FRamework EDitor) in this case. It is possible to program tables due to the user requirements under FRED. As it is further possible to record macro sequences, the calculations can be done fully automatically. The mentioned features are part of most of the commercially available table calculation programs.

The first Framework III table which is applied, stores all experimental data and calculates the swelling pressure–concentration curves. Their data are stored to disk as an ASCII file. Now, the basic program Flory is loaded which performs a non-linear numerical iteration of a slightly modified Flory-Huggins equation due to the Gauss-Jordan procedure. The Flory-Huggins equation is described elsewhere [3, 4]. The iteration yields the concentration dependent χ_w-interaction parameter in the mass fraction scale (index w) as the network constant C_w. These data are stored to disk in an ASCII file and read into a second Framework III table which calculates the elastic modulus, the shear modulus, and the number average molar mass of the elastically active network chains [3, 4].

Conclusions

It is possible to greatly improve the efficiency of ultracentrifugal experiments by introducing a modulation system with a laser light source modulated by means of an acusto-optical modulator (AOM). Such a modulated light source enables the use of multiplace rotors and multichannel cells. The described modulation system with modulation times of less than 1 μs ensures an angular resolution of less than 0.3° of the location of the light flashes on the rotor. Therefore, up to 14 sequential light flashes can be placed into the

4° sector of a monosector cell at the maximum rotor speed of 60 000 RPM. A modified eight-cell rotor has been introduced for use with the model E centrifuge which overcomes the present limitation of using up to five cells with a six-place rotor. Together with the 10-hole centerpieces which are described in the second part of this trilogy [14], it is possible to simultaneously investigate up to 70 samples under exactly the same conditions.

Such a sample number requires an automated picture evaluation process. To install a fully automatic picture digitization system, the intensity of the light source has to be increased significantly because the digitization system does not work with an exposure time like the former photographic system. The laser has to have a power of at least 40 mW, which requires an AOM with a sufficient power density of laser light.

The applied picture digitization system works with the relative low standard VGA resolution of 640 ∗ 480 pixel in 256 colors/gray scales which results in a maximum horizontal resolution of 25 μm/pixel. It could be shown that the 256 gray scales ensure a good picture quality. As the concentration gradient could not be detected in the whole gel phase of the samples which have been investigated up to now, the evaluation is only possible by means of a mass balance [7]. Therefore, only the positions of the phase boundaries in the beginning of the experiment and the equilibrium case as the location of the cell bottom have to be used for the picture evaluation. This means that only the horizontal positions are important for the evaluation of equilibrium experiments with gels. The error of 1 pixel in the horizontal resolution results in a negligible relative error in the calculated swelling pressure and a low relative error of, at maximum, 1.2% in the corresponding polymer concentration if 10-hole centerpieces are used. The error becomes much less if thermodynamic and elastic properties are calculated because a non-linear numerical iteration is used which minimizes the errors. Therefore, the error of one pixel in the picture evaluation process is negligible for the derived results. As sedimentation equilibrium experiments with gels are very sensitive and reproducible, it is sufficient to measure every sample only once [7]. The time needed to reach equilibrium in an experiment in a 10-hole centerpiece is about 2 days.

If the vertical position of the Schlieren pattern, namely, dn/dr should be evaluated, a frame grabber with a vertical resolution of at least 600 pixel is recommended for the applied digitization system due to the camera resolution of 625 pixel.

The picture digitization needs just a few seconds. After this, the 256 colors/gray scale picture is converted to a black/white picture which has the advantage to require less disk space and to display only the desired picture details. The conversion can be repeated until sufficient results are achieved. This process takes less than a minute if the picture has enough contrast and the user is already skilled.

After the series of black/white pictures for one experiment has been stored on the harddisk, it is read into a sequence of self-written basic programs and Framework III tables. The fully automatic picture conversion from the BMP to the more compact GIF Files takes about 5 1/2 minutes on the applied PC (386–40 MHz with a math coprocessor) for a simple experiment with an eight-hole rotor and 10-hole centerpieces with only two pictures per sample (start and equilibrium). The fully automatic picture evaluation needs about 17 min, whereas the calculation of the swelling pressure–concentration curves in a Framework III table needs 28 min. The time needed for the calculation of the interaction parameter and the network constant from the swelling pressure–concentration curve by the program Flory depends on the starting values. A typical time range for one of the 70 curves is 1/2–1 min. The evaluation of these parameters for all 70 samples in a Framework III table needs only seconds if the values are read at one time.

The described software delivers the following thermodynamic and elastic parameters characterizing the physically or chemically crosslinked gels according to [4, 7]:

- Swelling pressure in dependence of the polymer concentration in a concentration range depending on the applied rotational speed;
- concentration dependence of the Flory-Huggins interaction parameter χ_w in the mass fraction scale as the corresponding concentration terms for χ_w;
- network constant C_w in the Flory-Huggins equation [4];

- volume factor B in the Flory-Huggins equation if it should differ from 0 [7];
- number average molar mass of the elastically effective network chains;
- Young modulus E and static shear modulus G.

From these parameters information about the network structure (crosslinking density, degree of branching, homogeneity) can be derived [4, 7]. As the change of the chemical potential of the solvent inside the gel phase is accessible by the derived results, some other information like stability limits of the system are available [7]. With this information for up to 70 samples, which have been measured under identical conditions, it is possible to characterize a complete gel/solvent system in the experimental accessible range by a set of adequate parameters at the desired temperature. To the authors' knowledge this is not possible by means of any other method.

The only restriction is that the samples have to be transparent up to now. But if the adhesion of the gel at the cell walls and cell windows can be avoided nearly completely, the meniscus gel/solvent will be sharp. If the solvent is transparent it will be possible to evaluate experiments with non transparent gels by assuming that the detected phase boundary between the non transparent gel phase and the transparent sol phase is the meniscus gel/solvent.

It should be stated that supplementary measurements for the equilibrium experiments with gels may be extremely difficult and time-consuming. It may be possible that the supplementary measurements last longer than the ultracentrifugal experiment. It is not only a problem to measure the densities of gels [22, 28], but also to determine the polymer concentration of the maximum swollen gel if the swelling process of a thermoreversible gel is superimposed by a dissolution process [3, 4]. This means that the method described cannot work at its maximum efficiency if many different gel/solvent systems are investigated at many different temperatures where a solvation of the gels in an amount of pure solvent occurs.

Acknowledgements

The authors are very grateful to the Deutsche Forschungsgemeinschaft (DFG) for financial support of this work. We thank Mannesmann Demag AG Duisburg and especially Dipl. Ing. Mobers for help in designing and modifying the eight-hole titanium rotor for use in the Beckman model E centrifuge, as for counterbalancing of the rotor. Furthermore, we thank Mr. Kupperschmidt and the staff of the mechanical workshop of FB 6 of the Universität-GH-Duisburg for modifying the rotor. Dipl. Chem. H. Hinsken is acknowledged for careful reading of the manuscript.

References

1. Tanaka T (1987) In: Mark, Bikales, Overberger, Menges (eds) Encyclopedia of Polymer Science and Engineering. Vol 7, John Wiley & Sons, p 514
2. Cölfen H (1991) Diplomarbeit, Duisburg
3. Holtus G, Cölfen H, Borchard W (1991) Progr Colloid Polym Sci 86:92
4. Borchard W, Cölfen H (1992) Makromol Chem, Macromol Symp 61:143
5. Hinsken H (1992) Diplomarbeit, Duisburg
6. Hinsken H, Borchard W (1993) submitted
7. Cölfen H (1993) Dissertation, Duisburg
8. Holtus H (1990) Dissertation, Duisburg
9. Cölfen H, Borchard W (1991) Progr Colloid Polym Sci 86:102
10. Holtus G, Borchard W (1989) Colloid Polym Sci 267:1133
11. Yphantis DA (1960) "Rapid determination of molecular weights of peptides and proteins" Ann N Y Acad Sci 88:586
12. van Holde KE, Baldwin RL (1958) J Phys Chem 62:734
13. Laue TM (1992) "Short Column Sedimentation Equilibrium Analysis for Rapid Characterization of Macromolecules in Solution" Beckman Instruments, Technical Information
14. Cölfen H, Borchard W, "A modified experimental set-up for sedimentation equilibrium experiments with gels: Part II Technical developments" (submitted to Anal Biochem)
15. Cölfen H, Borchard W, "A modified experimental set-up for sedimentation equilibrium experiments with gels: Part III Results" (in preparation)
16. Borchard W (1991) Progr Colloid Polym Sci 86:84
17. Borchard W (1993) this issue
18. Svedberg T, Pedersen KO (1940) Die Ultrazentrifuge. Steinkopff, Dresden
19. Johnson P (1971) J Photograph Sci 19:49
20. Johnson P (1970) Velocity and equilibrium aspects of the sedimentation of agar gels in Photographic Gelatin I, Academic Press, London: 13
21. Bloomfield VA (1976) Biopolymers 15:1243
22. Cölfen H, Borchard W (1993) submitted to Makromol Chem
23. Paul CH, Yphantis DA (1972) Anal Biochem 48:588
24. Ortlepp B, Panke D (1991) Progr Colloid Polym Sci 86:57
25. Mächtle W, Klodwig U (1989) Colloid Polym Sci 267:1117
26. Laue TM (1992) "On-line Data Aquisition and Analysis from the Rayleigh Interferometer" in Harding SE et al (Ed) "Analytical Ultracentrifugation in Biochemistry and Polymer Science" Royal Society of Chemistry, Cambridge, England: 63

27. Laue TM, "Real time interferometry in the ultracentrifuge" (1993) SPIE Proceedings Vol 1895 "Ultrasensitive Laboratory Diagnostics" (in press)

28. Holtus G, Kiepen F, Schwark K, Borchard W (1990) Makromol Chem, Rapid Commun 11:177

Received June 18, 1993;
accepted November 29, 1993

Authors' address:

Prof. Dr. W. Borchard
Universität-GH-Duisburg
Angewandte Physikalische Chemie
Lotharstraße 1
47048 Duisburg, FRG

Progress in Colloid & Polymer Science   Progr Colloid Polym Sci 94:102–106 (1994)

Molar mass distribution of poly(methyl methacrylate) by sedimentation analysis in a theta solvent

A. Jungblut, B. Ortlepp, and D. Panke

Röhm GmbH, Darmstadt, FRG

Abstract: Molar mass distributions of poly(methyl methacrylate) samples with molar masses in the range between 50 000 and 20 000 000 $g \cdot mol^{-1}$ and with low polydispersities were measured by sedimentation analysis in the analytical ultracentrifuge (Beckman Model E) with UV optics ($\lambda = 226$ nm) using acetonitrile as solvent at the theta temperature (35 °C).

We found good agreement of the weight average molar masses M_w with those determined by light scattering and by size exclusion chromatography (SEC), respectively; however, there are somewhat larger discrepancies in the polydispersities (AUC vs. SEC). In contrast to literature data, we found that with 35 °C, we were slightly below the theta temperature, so that the coefficients in the relationship between sedimentation constant and molar mass had to be fitted with the light-scattering results.

Further experiments will aim at an improved signal/noise ratio by employing a very fast analogue/digital converter for the data acquisition.

Key words: Analytical ultracentrifuge – molar mass distribution – poly(methyl-methacrylate) – sedimentation analysis – theta system – UV scanner – light scattering – size exclusion chromatography

Introduction

According to our company's product lines, we have mainly to deal with the characterization of poly(acrylates) and poly(methacrylates). In the case of ionomers and of very high molar mass samples the analytical ultracentrifuge (AUC) very often is the only tool for getting molar mass distributions. Extrapolation of the experimental data to infinite dilution may yield considerable errors unless the concentrations are comparatively small; to use UV optics with small wavelengths (about 220 to 230 nm) seems to be the best choice.

We wanted to characterize a series of poly(methyl methacrylate) (PMMA) samples by sedimentation velocity in the AUC with the aim to compare weight average molar masses M_w with light scattering (LS) results and molar mass distributions with size exclusion chromatography (SEC) measurements, respectively. This work was integrated in the development of an improved UV

system for one of our Beckman model E AUCs [1, 2] and should demonstrate what has been achieved so far and which further improvements will have to be done.

Experimental

In order to minimize extrapolation errors, we used acetonitrile as solvent at 35 °C (theta condition where interactions between the molecules are minimized), concentrations as small as possible (0.1–1.0 g/l), and a wavelength of 226 nm. Most of the samples had been prepared by anionic polymerization and have molar masses between $5 \cdot 10^4$ and $2 \cdot 10^7$ $g \cdot mol^{-1}$; they are routinely used for SEC calibration.

Figure 1 shows the optical pathways of the centrifuge with Schlieren and UV optics, the latter one being the subject of this presentation.

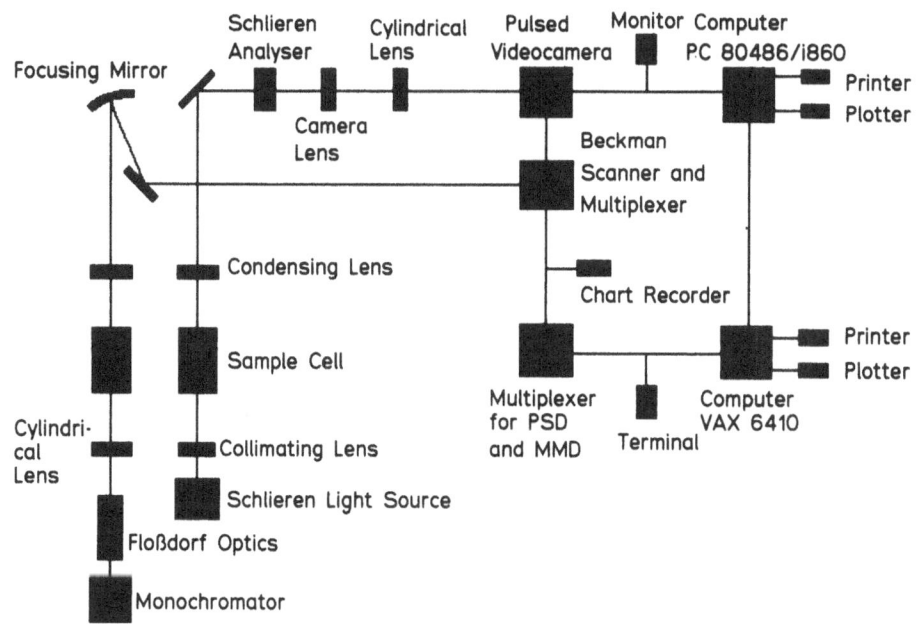

Fig. 1. Optical and electronic pathways of the AUC system with Schlieren optics and with UV photoelectric scanner

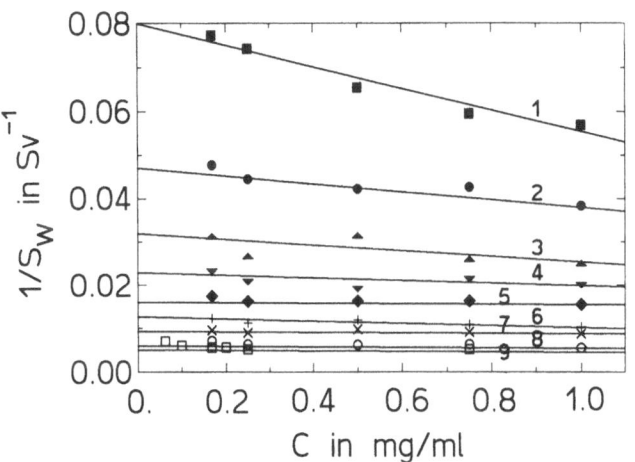

Fig. 2. The concentration dependence of the reciprocal sedimentation coefficients for PMMA samples 1 to 9 (see Table 1)

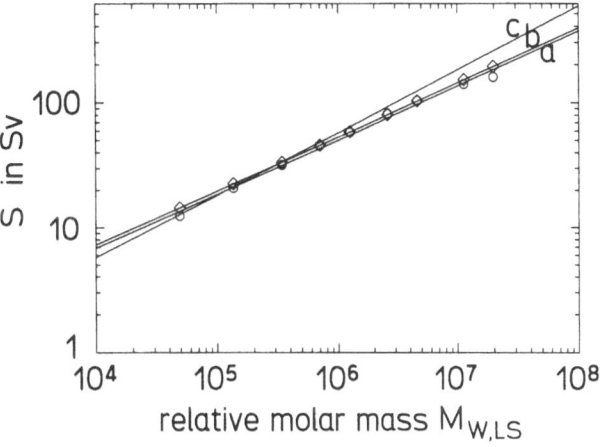

Fig. 3. Sedimentation coefficient molar mass relationship. a) Circles: experimental data, $S = 0.127 \cdot M^{0.433}$; b) diamonds: after an optimization procedure (see text), $S = 0.137 \cdot M^{0.432}$; c) line without symbols: results by Härzschel [4], $S = 0.058 \cdot M^{0.50}$

Results

Figure 2 gives for our system, viz. PMMA/acetonitrile at 35 °C, the concentration dependence of the reciprocal sedimentation coefficient S_w. The S_w values were taken from the $G(S)$ distributions as their first moments. The range of concentrations is about one order of magnitude lower then concentrations used, for example, by Boog et al. [3] for the same system.

Extrapolation to $c = 0$ according to

$$\frac{1}{S_w} = \frac{1}{S_{w0}} (1 + k_s c) \tag{1}$$

yields for each sample S_{w0} and k_s. Together with the absolute molar masses M_w obtained by light scattering the $M-S$-relationship

$$S_{w0} = K \cdot M^a \tag{2}$$

with $K = 0.127$ and $a = 0.433$ can be established. It is shown in Fig. 3 together with results by Härzschel [4]. The differences may be explained by slightly different solvent qualities and/or temperatures. Furthermore, we show our results modified by an optimization procedure which will be explained briefly.

We use the $M-S$-relationship from the first moments of the $G(S)$ distribution only as a first approximation for computing the molar mass distribution from the $G(S)$ distributions. Only if K and a are correct will the weight average molar masses calculated from these molar mass distributions match the light-scattering results. So, the weight average molar masses of the molar mass distributions are compared with the light-scattering M_w values and, using a search method, the coefficients K and a are fitted until

$$\Sigma(M_{w,LS} - M_{w,AUC})^2$$

becomes a minimum. We find the relationship $S_0 = 0.137 \cdot M^{0.432}$. Fortunately, curves a and b in Fig. 3 are almost identical.

The MMDs obtained by this procedure are together with SEC distributions depicted in Fig. 4 and, for the two samples with the highest molar masses, in Fig. 5. Here, no SEC results are accessible because even the best SEC columns available have an exclusion limit at molar masses of some 10^6 g·mol^{-1}. Some of the AUC distributions are most certainly capable of improvement which will be discussed later. However, the agreement of the M_w values with light scattering results is quite

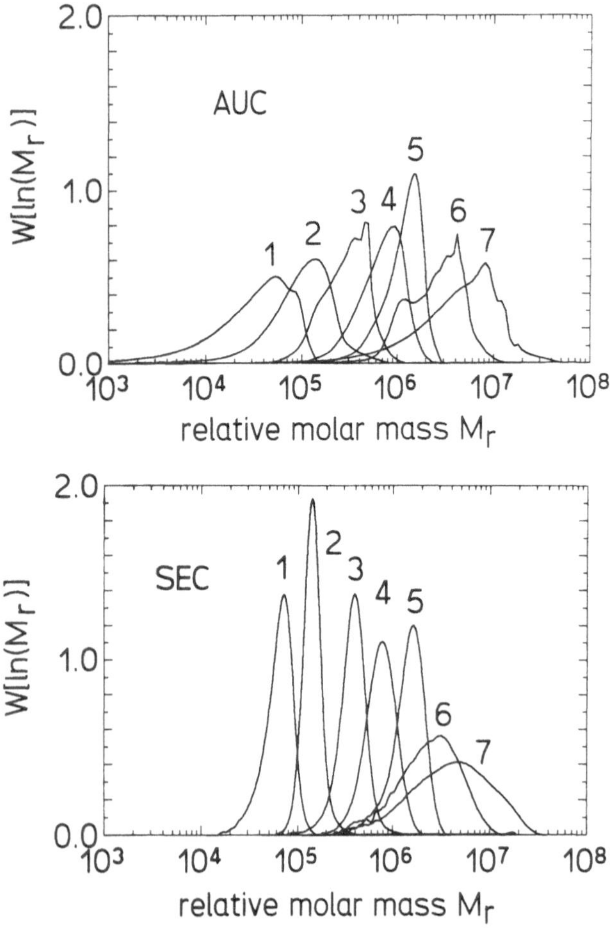

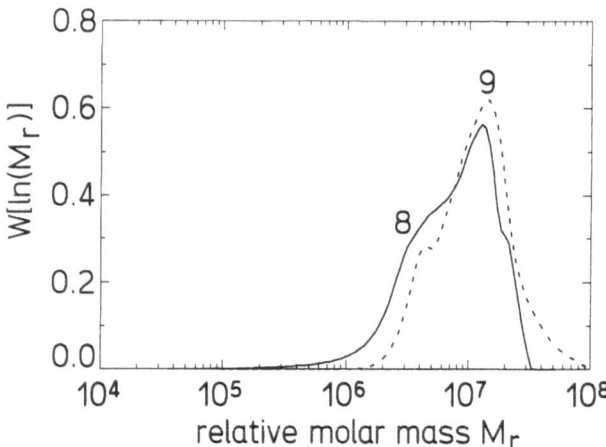

Fig. 5. AUC molar mass distributions for two high molar mass samples. $8: M_{w,LS} = 11.3 \cdot 10^6$ g·mol^{-1}, 9: $M_{w,LS} = 19.8 \cdot 10^6$ g·mol^{-1}

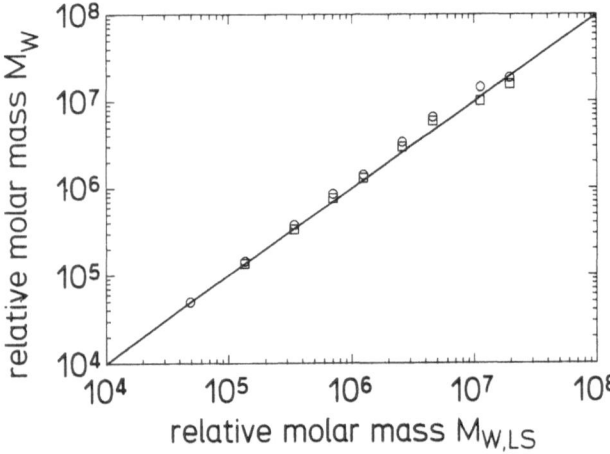

Fig. 6. M_w results from AUC versus light-scattering data $M_{w,LS}$. Circles: without optimization; squares: with optimization (see text)

Fig. 4. Molar mass distributions of samples 1 to 7 (see Table 1). a) AUC measurements; b) SEC results

good (see Fig. 6). This probably is the benefit of the theta system which provides for comparatively small k_s values (see Eq. (1) in contrast to a good solvent like, for example, acetone, see Fig. 7.

The results of LS, AUC, and SEC measurements are shown in Table 1.

As a last example, we show in Fig. 8 results obtained with a mixture of two samples. The MMD is compared with a superposition of the two individual MMDs (curves 2 and 5 in Fig. 4a). Although the correct mass ratio is found, the influence of the Johnston–Ogston effect [5] can be clearly seen: in spite of the very low concentrations the smaller molecules are accelerated by the large ones which, in turn, are slowed down by the small ones. Apparently, the Johnston–Ogston ef-

fect cannot be eliminated even in the theta system or by the very small concentrations used.

Discussion and outlook

A comparison of the molar mass data (see Table 1) shows good agreement between the AUC and the LS results. This has to be expected since the light-scattering data had been used for fitting the constants K and a in the M–S-relationship. In addition, it is not surprising that the agreement with the SEC data is quite fair, too, because our PMMA samples had been used (among others) for establishing the SEC calibration curve. In

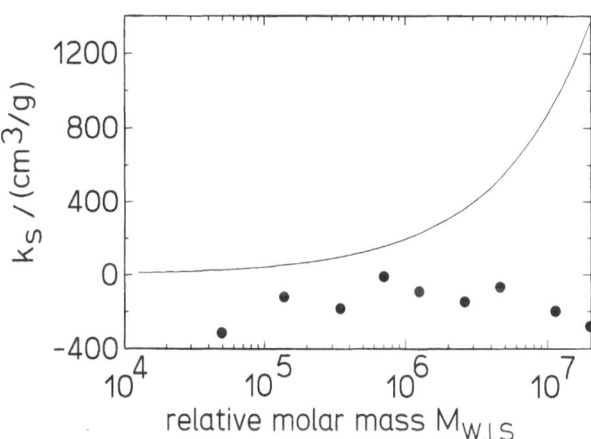
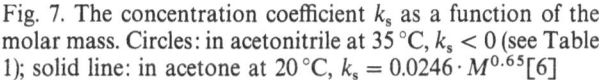

Fig. 7. The concentration coefficient k_s as a function of the molar mass. Circles: in acetonitrile at 35 °C, $k_s < 0$ (see Table 1); solid line: in acetone at 20 °C, $k_s = 0.0246 \cdot M^{0.65}$ [6]

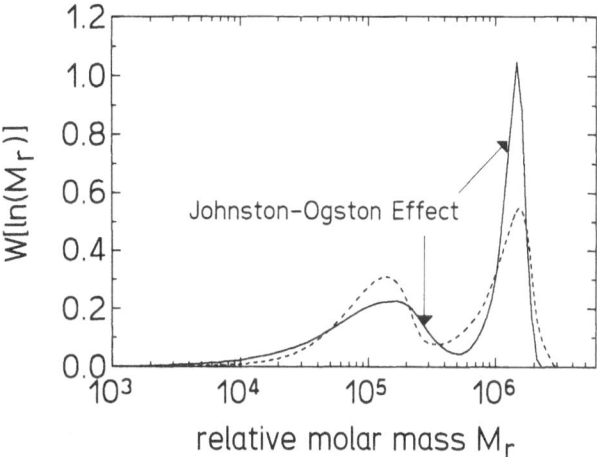

Fig. 8. Molar mass distribution of a 1:1 mixture of two samples. Solid line: experimental MMD of the mixture (mass ratio found 51:49); broken line: superposition of the two individual experimental MMDs (curves 2 and 5 in Fig. 4a)

Table 1. Molar masses M_w (in g/mol), heterogeneity indices $U(= M_w/M_n - 1)$, experimental sedimentation coefficients S_{w0} (in Sv), sedimentation coefficients S_0 after optimization, and concentration coefficients k_s (in cm^3/g)

	LS	AUC					SEC	
Sample code	M_w	M_w	U	S_{w0}	S_0	k_s	M_w	U
3835/81/N	49,100	46,000	1.37	12.5 ± 0.1	14.5	− 314.1	66,500	0.15
5911/04/E	137,000	136,000	0.72	20.8 ± 0.6	22.6	− 120.9	146,000	0.06
5911/04/N	342,000	336,000	0.32	31.9 ± 2.3	33.6	− 182.8	398,000	0.13
3835/93/S	702,000	763,000	0.34	45.6 ± 2.8	45.8	− 9.8	790,000	0.16
5911/05/R	1,250,000	1,319,000	0.32	58.4 ± 1.5	58.8	− 92.0	1,550,000	0.23
3835/93/R	2,600,000	2,967,000	0.55	81.6 ± 2.8	80.6	− 146.6	2,990,000	0.69
8844/17	4,620,000	5,946,000	0.66	104.0 ± 4	103.0	− 65.3	5,900,000	1.33
8844/22	11,300,000	10,200,000	0.43	140.0 ± 6	152.0	− 196.7		
7825/08	19,800,000	15,870,000	0.51	159.0 ± 9	194.0	− 278.9		

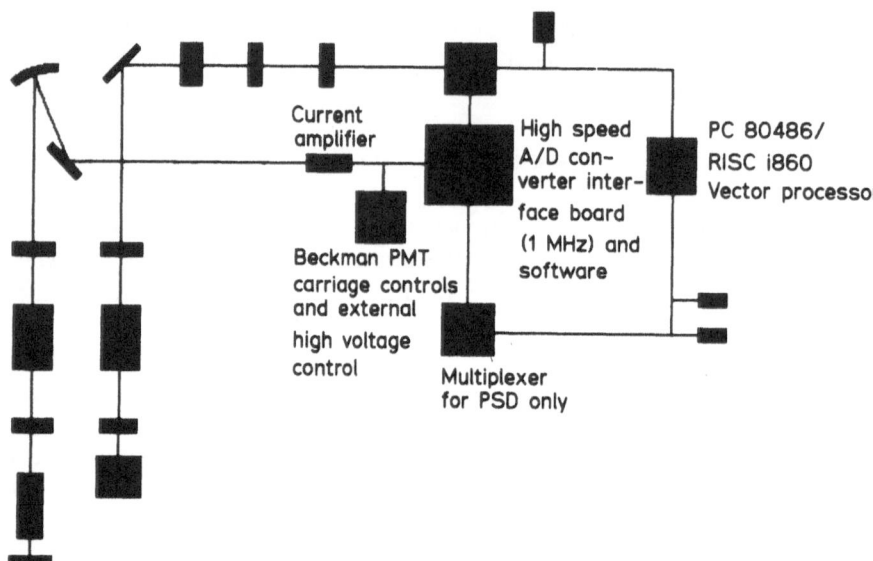

Fig. 9. Planned improvement of the signal processing in the AUC system (see Fig. 1 for the present state)

most cases, however, the AUC mass distributions are broader then those obtained by SEC. The SEC heterogeneity indices U should be quite reliable except for very high molar mass samples where exclusion effects must be expected. The Johnston–Ogston effect should narrow the molar mass distributions, but a look at Fig. 4 hints that, on the contrary, there is a low molar mass tailing in the AUC results. The reason for this is not yet clear, but it has to be checked whether there are problems with the baseline, with the linearity of the photomultiplier, or with the numerical extrapolation procedures.

Some problems of the molar mass distributions are probably due to our present hardware in connection with the very low polymer concentrations. In order to improve the signal/noise ratio the following measures will be taken:

– The original Beckman electronic parts in the UV scanner which are prone to trouble will be replaced by external parts, viz. a preamplifier and a high voltage control for the photomultiplier; this will improve the S/N ratio by a factor of 10 at least. Furthermore, a fast real time A/D converter (see Fig. 9) with a sampling rate of 1 MHz will substantially decrease the time for data processing in conjunction with homemade software for acquisition and evaluation of the fast signals.

– The raw data will be smoothed by a Bezier polynome fit. This is very time-consuming and has to be done by a RISC i860 vector processor.

This, hopefully, will give us the possibility to measure with good accuracy MMDs of (meth) acrylate polymers, although their UV absorbancies are relatively low.

References

1. Ortlepp B, Panke D (1991) Progr Colloid Polym Sci 86:57–61
2. Ortlepp B, Panke D (1992) Makromol Chem, Makromol Symp 61:176–184
3. Boog U, Tianbao H, Meyerhoff G (1987) Eur Polym J 23:781–785
4. Härzschel R (1985) Dissertation Mainz
5. Johnston JP, Ogston AG (1946) Trans Faraday Soc 42:789–796
6. Stickler M (1977) Dissertation Mainz

Received June 11, 1993
accepted October 26, 1993

Authors' address:

Dr. Dietrich Panke
Röhm GmbH, Abt. AP2–PC
64275 Darmstadt, FRG

Progress in Colloid & Polymer Science   Progr Colloid Polym Sci 94:107–115 (1994)

Sedimentation and diffusion of polyelectrolytes

P. M. Budd

Department of Chemistry, University of Manchester, Manchester, United Kingdom

Abstract: The sedimentation and diffusion behaviour of polyelectrolytes is discussed. The properties of polyelectrolyte solutions are profoundly influenced by intramolecular and intermolecular charge interactions. In the presence of a simple salt, the range of these interactions is reduced, but additional effects arise. For example, the Donnan effect needs to be taken into account in the quantitative analysis of sedimentation data. The Donnan parameter, Γ, may be estimated from the concentration dependence of sedimentation velocity. In the absence of a simple salt, sedimentation coefficients are rather low and diffusion coefficients, at moderate polymer concentrations, are rather high; this has been attributed to coupled motion of polyions and counterions. Results are presented from a synthetic boundary experiment on sodium poly(styrene sulphonate) in salt-free solution which show that the diffusion coefficient depends markedly on polymer concentration below $0.2 \, \mathrm{g \, dm^{-3}}$. This diffusion process appears to be equivalent to the fast diffusion process observed by dynamic light scattering. Current theories of coupled-ion diffusion in salt-free polyelectrolyte solutions are inadequate; the effects of intermolecular interactions and the structure of polyelectrolyte solutions need to be considered.

Key words: Polyelectrolyte – sedimentation velocity – diffusion – analytical ultracentrifuge – dynamic light scattering – sodium poly(styrene sulphonate)

Introduction

Polyelectrolytes exhibit a combination of polymeric and electrolyte behaviour that gives rise to many unique and useful properties; they are utilized as viscosity modifiers, flocculants, dispersants, water softeners and in many other applications. A great deal of useful information about such materials can be obtained by sedimentation and diffusion measurements [1–3]. This paper discusses the sedimentation velocity and diffusion behaviour of polyelectrolytes, indicating the limits of current knowledge and some of the problems which still remain to be solved.

Polyelectrolytes

Ionization of a polyelectrolyte in aqueous solution gives rise to a multiply-charged polyion and many small counterions (Fig. 1). Interactions between charged species have a profound effect on the properties of polyelectrolyte solutions and are particularly significant in the absence of additional low molecular weight electrolyte.

Intramolecular interactions between charges on a polyion may influence the conformation of the polymer. It is generally thought that polyelectrolyte chains are more highly expanded in solution than would be expected for an equivalent nonionic polymer. Polyelectrolytes are often treated as being rod-like at low concentration in pure water, though recent evidence has cast some doubt on the extent to which polyion chains are expanded [4–7].

Interactions between a polyion and its counterions restrict the freedom of those counterions, so that their activity coefficients may be very low. For a high charge density polyelectrolyte some counterions may be "condensed" onto, or bind

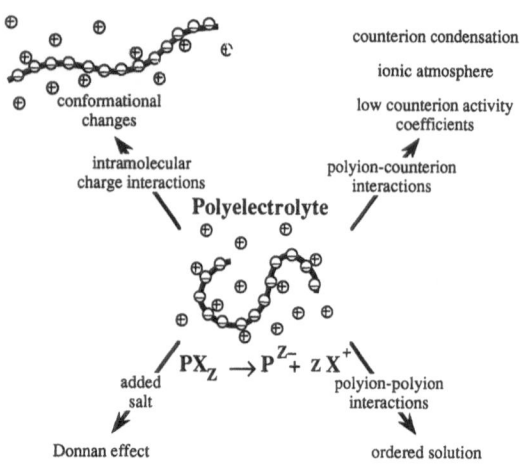

Fig. 1. Illustration of aspects of polyelectrolyte behaviour

specifically with, the polyion, whilst others are loosely associated in an "ionic atmosphere".

Intermolecular interactions between polyions may give rise to a degree of order even in very dilute solutions [8]. There is increasing evidence that this may be more significant than has previously been appreciated.

The range of each of these interactions is reduced by addition of a low molecular weight electrolyte. In many respects, a polyelectrolyte in the presence of a simple salt behaves much like a nonionic polymer. However, the presence of a third component can itself give rise to various complications, such as the Donnan effect outlined below.

The Donnan effect

The Donnan effect may be studied in a membrane distribution experiment, in which a polyelectrolyte solution is separated from a solution of a simple salt by a semi-permeable membrane. The polyion cannot pass through the membrane but the counterions and the constituent ions of the simple salt can. At equilibrium, there is a lower concentration of salt in the presence of polyion than in its absence; in other words salt is excluded from the polyelectrolyte solution to some extent [9]. This may be described in terms of a Donnan membrane distribution parameter, Γ, which is defined at infinite dilution of the polyelectrolyte to avoid the complication of

interactions between polyions:

$$\Gamma = \lim_{C_u \to 0} \left(\frac{C_s' - C_s}{C_u} \right), \tag{1}$$

where C_s' and C_s are the concentrations (in mol dm^{-3}) of salt on the polyion-free and polyion-containing sides of the membrane respectively, and C_u is the concentration of polyelectrolyte per charged unit. It can be shown that for the 'ideal' situation, in which the mean ionic activity coefficient for the salt is unaffected by the polyion, the value of Γ is 0.5. In reality, Γ is usually rather less than 0.5, but increases with increasing salt concentration.

The Donnan effect manifests itself wherever there is a concentration gradient of polyelectrolyte, as in a sedimentation or diffusion experiment. Furthermore, Donnan equilibrium represents a situation of constant chemical potential of diffusible components, which defines an appropriate density increment for use in the sedimentation analysis of polyelectrolytes.

In the sedimentation analysis of a multicomponent system such as a polyelectrolyte with added salt, it is advantageous to utilize a density increment, $(d\rho/dc)$, rather than a buoyancy factor, $(1 - \bar{v}\rho)$. This is important, for example, in the absolute determination of molecular weight by sedimentation equilibrium. It can be shown [10–12] that an appropriate increment for polyelectrolyte studies is that for the condition of constant chemical potential of diffusible components, $(d\rho/dc)_\mu$. This may be determined from measurements of the density, ρ, of solutions with different polymer concentrations, c, which have been brought into dialysis (Donnan) equilibrium with a salt solution. The increment $(d\rho/dc)\mu$ differs from that for the condition of constant salt concentration, $(d\rho/dc)C_s$, because of the Donnan effect. If the salt has an ion in common with the polyelectrolyte, they are approximately related by

$$\left(\frac{d\rho}{dc} \right)_\mu \approx \left(\frac{d\rho}{dc} \right)_{C_s} - (1 - \bar{v}_s \rho) \left\{ \frac{M_s}{M_u} \Gamma \right\}, \tag{2}$$

where M_u is the molecular weight per charged unit of the polyelectrolyte and M_s and $\bar{v}_s$ are the molar mass and partial specific volume respectively of the salt. Note that as Γ generally increases with increasing salt concentration, the difference between the two increments generally increases with increasing salt concentration.

It is appropriate to ask how large an error might be incurred if the increment $(d\rho/dc)_{C_s}$ is used instead of $(d\rho/dc)_\mu$ in sedimentation analysis. For a sample of sodium poly(styrene sulphonate) in 0.5 M aq. NaCl the former increment has been measured as 0.464 ± 0.01 and the latter as 0.376 ± 0.01, a difference of about 20%. For other systems the difference may be less marked. Inspection of Eq. (2) reveals that the difference between the two increments should disappear for the specific case of a salt for which $(1 - \bar{v}_s\rho)$ is zero. A salt for which the buoyancy factor is close to zero in water at room temperature is tetramethylammonium chloride. For a sample of tetramethylammonium poly(styrene sulphonate) in 0.5 M aq. $(CH_3)_4NCl$, $(d\rho/dc)_{C_s}$ was measured as 0.210 ± 0.01 and $(d\rho/dc)_\mu$ as 0.203 ± 0.01 [13].

It is not always feasible to carry out equilibrium dialysis or to use a salt for which the buoyancy factor is minimal. In such cases it may be possible to evaluate Γ by some other technique and hence calculate $(d\rho/dc)_\mu$ from $(d\rho/dc)_{C_s}$ using Eq. (2). It will be seen below that it is possible to estimate Γ from the concentration dependence of sedimentation velocity.

Sedimentation velocity

The sedimentation velocity of a polyelectrolyte is strongly influenced by salt concentration [14]. This is illustrated in Fig. 2, which shows the variation of the reciprocal sedimentation coefficient, $1/S$, with polymer concentration at various ionic strengths for a sample of sodium poly(styrene sulphonate). In the absence of added salt, the sedimentation coefficient is extremely low (i.e. $1/S$ is large). A small increase in ionic strength gives rise to a considerable increase in sedimentation coefficient. This has been called the "primary charge effect" and has been attributed to the electric field which would be set up if a rapidly sedimenting polyion were separated from slowly sedimenting counterions [15]. Polyion and counterions are forced to move together at a relatively low velocity.

There have been a number of theoretical treatments of sedimentation in polyelectrolyte solutions. The most rigorous approach involves the

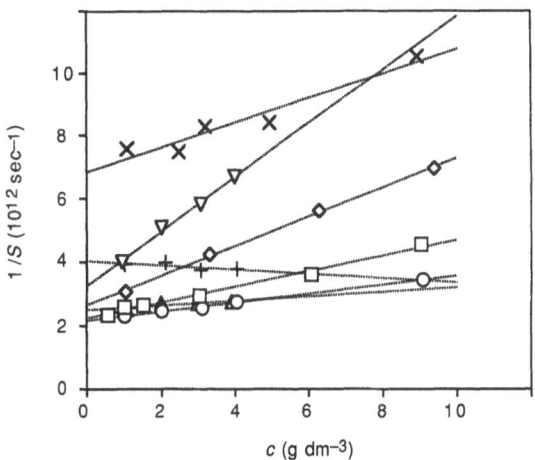

Fig. 2. Dependence of reciprocal sedimentation coefficient on polymer concentration for aqueous solutions of sodium poly(styrene sulphonate) ($M = 1.0 \times 10^5$) with no added salt ($\times$), and with 0.005 mol dm^{-3} NaCl ($\bigtriangledown$), 0.02 mol dm^{-3} NaCl($\diamond$), 0.1 mol dm^{-3} NaCl ($\square$), 0.5 mol dm^{-3} NaCl($\bigcirc$), 2.0 mol dm^{-3} NaCl($\triangle$) and 4.17 mol dm^{-3} NaCl($+$)

application of irreversible thermodynamics [16, 17]. The classical approach considers the actual velocity of a polyion as having two components; there is a contribution from the centrifugal field and a contribution from the internally created electric field which would arise if the polyion were separated from its counterions [11, 12, 18].

Sedimentation of polyelectrolyte in salt-free solution

For a polyelectrolyte PX_z in a salt-free solution the limiting sedimentation coefficient, S_0, is given by [12]

$$S_0 = \frac{M(1 - \bar{v}\rho)}{f + iZf_x}, \tag{3}$$

where M is the molecular weight and $\bar{v}$ is the partial specific volume of the polyelectrolyte, ρ is the density of the solution, f and f_x are the molar translational friction coefficients of the polyion P and counterion X respectively and i is an effective charge parameter which takes account of interactions between the polyion and its counterions. It is assumed that, on average, $(1 - i)Z$ counterions interact in some way with the polyion so that it carries an effective charge of iZ. The magnitude of i depends both on the number of any condensed counterions and on the extent of screening by the ionic atmosphere. Equation (3) predicts that the limiting sedimentation coefficient will be rather low for a polyelectrolyte in the absence of added salt, as is observed experimentally.

Sedimentation of polyelectrolyte in the presence of simple salt

For a polyelectrolyte PX_z in the presence of a simple salt, S_0 is given in the absence of secondary charge effects by [12]

$$S_0 = \frac{M(1 - \bar{v}\rho) - 0.5iZM_s(1 - \bar{v}_s\rho)}{f} . \quad (4)$$

The Donnan parameter Γ may, in equilibrium situations, be related to an effective charge parameter i by

$$\Gamma = 0.5i . \quad (5)$$

If the same identification is made in the case of sedimentation velocity, the equation for S_0 becomes

$$S_0 = \frac{M(d\rho/dc)_\mu}{f} , \quad (6)$$

which shows that in some respects one can treat a polyelectrolyte with added salt similarly to a nonionic polymer, provided that the density increment for the condition of Donnan equilibrium is used. This equation may be combined with the corresponding equation for the limiting diffusion coefficient (see below) and a molecular weight evaluated. For a more detailed analysis of sedimentation behaviour, however, one should bear in mind that both $(d\rho/dc)_\mu$ and f vary with salt concentration, reflecting variations in Γ, in the conformation of the chain and in the thickness of the ionic atmosphere.

Concentration dependence of S

For a polyelectrolyte in the presence of a simple salt, $1/S$ is linearly dependent on the polymer concentration and the slope of such a plot decreases with increasing salt concentration, as can be seen in Fig. 2. According to the classical theory as developed by Eisenberg, [12] charge effects give rise to a slope, k, given by

$$k \approx \frac{10^3 i^2 f_Y}{M_u C_s[M_u(1 - \bar{v}\rho)(1 + f_Y f_X^{-1}) - iM_s(1 - \bar{v}_s\rho)]}$$

$$\times \left\{ 1 - \frac{M_s(1 - \bar{v}_s\rho)}{f_X(1 + f_Y f_X^{-1})S_0} \right\} , \quad (7)$$

where f_Y is the molar translational friction coefficient of the coion. Rearrangement of this equation enables i to be calculated, as [13, 14]

$$\alpha i^2 + \beta i + \gamma = 0 , \quad (8)$$

where

$$\alpha = \frac{10^3}{kM_u C_s} \left\{ f_Y - \frac{M_s(1 - \bar{v}_s\rho)f_Y}{f_X(1 + f_Y f_X^{-1})S_0} \right\}$$

$$\beta = M_s(1 - \bar{v}_s\rho)$$

$$\gamma = - M_u(1 - \bar{v}\rho)(1 + f_Y f_X^{-1}) .$$

The problem remains, however, of isolating the concentration dependence due to charge effects from the normal concentration dependence exhibited by any polymer. To a first approximation we may assume that the experimental concentration dependence is

$$k_e = k_{ni} + k , \quad (9)$$

where k_{ni} is the slope expected for an equivalent nonionic polymer. For many flexible, nonionic polymers in good solvents it has been found that [19]

$$k_{ni} = \frac{1.6[\eta]}{S_0} , \quad (10)$$

where $[\eta]$ is the intrinsic viscosity (limiting viscosity number).

Since i may be equated to twice the Donnan parameter, Γ, Eqs. (8)–(10) provide a means of estimating Γ. For a sample of sodium poly(styrene sulphonate) $(\bar{M}_w = 1.01 \times 10^6)$ in 0.5 M aq. NaCl, Γ has been estimated in this way as 0.41. If this value is incorporated into Eq. (2), one gets $(d\rho/dc)_\mu = 0.385$, in good agreement with the measured value of 0.376 ± 0.01 [13].

Diffusion

The diffusion behaviour of a polyelectrolyte, like its sedimentation behaviour, is profoundly influenced by charge effects.

Diffusion of polyelectrolyte in the presence of simple salt

For a polyelectrolyte in the presence of a simple salt at constant chemical potential of the salt (i.e. for a boundary between a polyelectrolyte solution and a salt solution with which it is in Donnan equilibrium), the classical theory predicts [12]

$$\frac{1}{D}\frac{d\mu}{d\ln C} = f + \frac{i^2 Z^2 C}{(f_X^{-1} + f_Y^{-1})C_S + iZf_X^{-1}C}, \quad (11)$$

where D is the mutual diffusion coefficient. At infinite dilution of the polymer

$$D_0 = \frac{RT}{f}, \quad (12)$$

which may be combined with the equivalent expression for S_0 to enable molecular weights to be determined.

Diffusion of polyelectrolyte in salt-free solution

In the absence of a simple salt, observed diffusion coefficients for polyelectrolytes at finite polymer concentrations are remarkably high. This has in the past been attributed to Pederson's "primary charge effect", alternatively termed "coupled-ion diffusion". It is argued that the highly mobile counterions accelerate diffusion of the macromolecule [12, 20]. If this explanation is correct, one would expect a high diffusion coefficient at infinite dilution of the polymer.

The classical theory of the primary charge effect as developed by Eisenberg [12] predicts, for a polyelectrolyte in salt-free solution

$$D_0 = \frac{RT}{f + iZf_X}. \quad (13)$$

In other words, Eisenberg predicts that the limiting diffusion coefficient, like the limiting sedimentation coefficient, should be rather low. However, other workers have obtained expressions which imply that D_0 should be magnified by a factor $(1 + iZ)$ [17, 20, 21]. This discrepancy between different theoretical treatments warrants further investigation.

It was observed long ago that a polyelectrolyte in salt-free solution exhibits a characteristic sharp diffusion front [22, 23]. Figure 3 shows results from a diffusion experiment carried out in an analytical ultracentrifuge. In this experiment, a sharp boundary was established between a carefully purified solution of sodium poly(styrene sulphonate) and pure water with the aid of a capillary-type synthetic boundary centrepiece. The spreading of the boundary was observed with a Rayleigh interference optical system. A low rotor speed was employed so that sedimentation was not significant; the analytical ultracentrifuge merely provides a means of forming the initial boundary and a convenient optical system for following the spreading of that boundary. An anomaly is seen at low concentrations in the concentration/position curve (Fig. 3a). This is accentuated when the derivative is taken (Fig. 3b), as would be observed with a Schlieren optical system.

These data may be interpreted in terms of a diffusion coefficient that depends markedly on concentration. The equation for one-dimensional diffusion in the x direction is [23, 24]

$$\frac{\partial c}{\partial t} = \frac{\partial}{\partial x}\left(D\frac{\partial c}{\partial x}\right). \quad (14)$$

A century ago, Boltzmann [25] showed that by introducing a new variable, $\eta = x/2t^{1/2}$, an ordinary differential equation could be obtained:

$$-2\eta\frac{dc}{d\eta} = \frac{d}{d\eta}\left(D\frac{dc}{d\eta}\right). \quad (15)$$

Equation (15) may be solved to enable D to be calculated. For an experiment with initial conditions

$$t = 0, \quad c = c_\infty, \quad x < 0$$
$$t = 0, \quad c = 0, \quad x > 0;$$

the diffusion coefficient at any concentration c_1 between 0 and c_∞ is given by [24]

$$D_{c=c_1} = -\frac{1}{2t}\frac{dx}{dc}\int_0^{c_1} x\,dc. \quad (16)$$

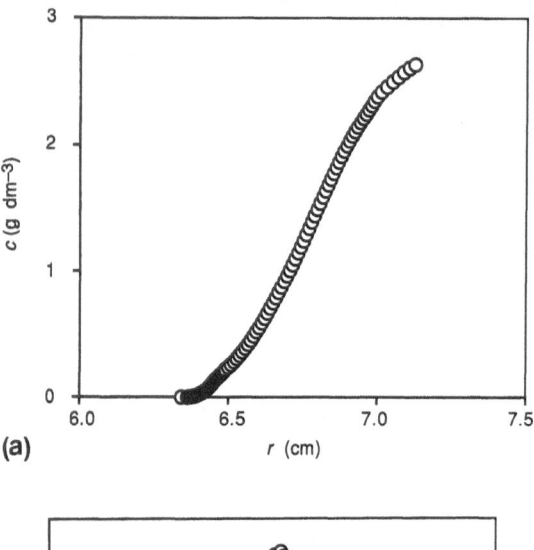

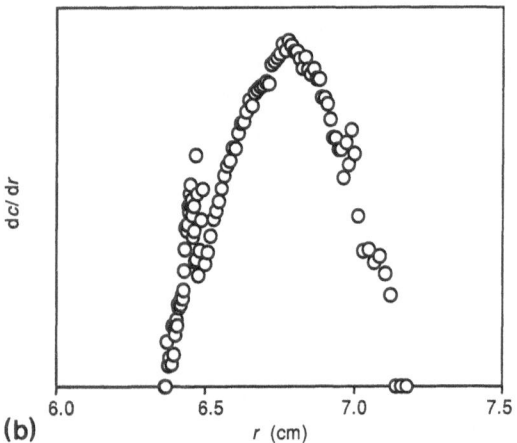

Fig. 3. Results from a synthetic boundary diffusion experiment in an analytical ultracentrifuge cell; a) concentration curve, b) concentration gradient curve [purified sodium poly(styrene sulphonate) ($M = 1.06 \times 10^6$) in water, initial concentration: $2.7\ \mathrm{g\,dm^{-3}}$; time after formation of initial boundary: 4440 s]

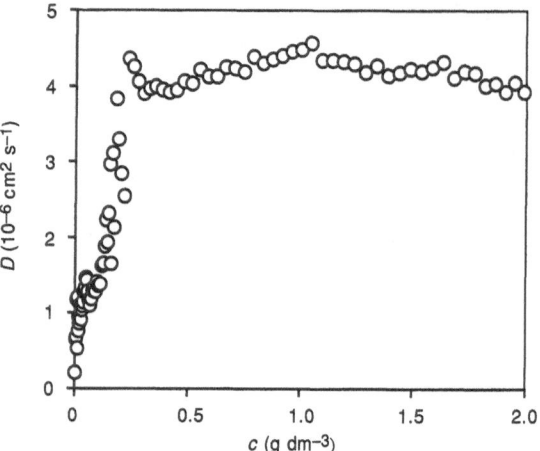

Fig. 4. Dependence of diffusion coefficient on polymer concentration for sodium poly(styrene sulphonate) ($M = 1.06 \times 10^6$) in salt-free solution, determined from synthetic boundary data by the Boltzmann–Matano procedure

Equation (16) enables the diffusion coefficient/concentration curve to be calculated over the whole concentration range of the boundary, using the results at a single, finite time after formation of the initial boundary. It is necessary that x is measured from the initial boundary position, so that $\int x\,dc$ over the whole boundary is zero (i.e. mass is conserved in a constant volume system). This data analysis method was used by Matano [26] to determine diffusion coefficients in metal systems and is sometimes referred to as the Boltzmann–Matano procedure.

Equation (16) should not strictly be applied to an experiment carried out in a sector-shaped ultracentrifuge cell, where diffusion is not truly one-dimensional. However, we shall use it to give a semi-quantitative insight into the diffusion behaviour.

Figure 4 presents results calculated using Eq. (16) from the data of Fig. 3 for sodium poly(styrene sulphonate) ($M = 1.06 \times 10^6$). It can be seen in Fig. 4 that D is low at very low concentrations, but in the very dilute solution regime D increases rapidly with increasing concentration. At concentrations above about $0.2\ \mathrm{g\,dm^{-3}}$, D has a constant, high value; much higher than one would observe for a nonionic polymer of similar molecular weight. Few studies of the diffusion behaviour of polyelectrolytes in salt-free solution have been extended to sufficiently low concentrations to see the dramatic change in D.

Dynamic light scattering

It is of interest to compare the above results, from a synthetic boundary experiment in an ultracentrifuge cell, with apparent diffusion coefficients determined by dynamic light scattering. Dynamic light scattering from polyelectrolyte solutions at very low salt concentrations is far from straightforward; scattered light intensities are extremely low and correlation functions exhibit considerable deviations from single exponentiality [27].

Early studies appeared somewhat contradictory. Schurr et. al. [28] reported a sudden drop in the apparent rate of diffusion of poly(L-lysine) on decreasing the ionic strength. They referred to this as an "ordinary-extraordinary transition". Later work indicated that there were two diffusive processes occuring in the "extraordinary" region; a fast process and a slow process [29–34].

Recent studies by dynamic light scattering of quaternized poly(2-vinyl pyridine) [31] and sodium poly(styrene sulphonate) [32–34] are beginning to give a consistent picture. In salt-free solution (or as salt-free as is experimentally achievable) at the lowest polymer concentration for which measurements can be made, only one diffusion coefficient is obtained. As the polymer concentration is increased, two diffusive processes become apparent; the fast diffusion coefficient increases and the slow diffusion coefficient decreases as polymer concentration increases. Above a critical concentration the fast diffusion coefficient is essentially independent of polymer concentration, whilst the slow diffusion coefficient decreases.

Figure 5 compares the results obtained by the synthetic boundary method with results from dynamic light scattering [34] for a sample of sodium poly(styrene sulphonate) of similar molecular weight. Remarkable agreement is obtained between the diffusion coefficient determined from the spreading boundary and the *fast* diffusion coefficient determined by dynamic light scattering, given both the neglect of the sector-shape of the cell in the analysis of the synthetic boundary data and the difficulty of carrying out light scattering experiments under these conditions. It should be remembered that the synthetic boundary results come from a single experiment in which the shape of the boundary is analysed at a single time after its formation.

The results in Fig. 5 clearly suggest that the *fast* diffusion coefficient determined by dynamic light scattering is equivalent to that measured from the spreading of a macroscopic boundary. Dynamic light scattering results indicate that the fast diffusion coefficient is more or less independent of molecular weight [33]. The interpretation of this diffusion coefficient is open to dispute. Sedlák [30,33,34] has attributed it to coupled diffusion of polyions and counterions (i.e. the primary charge effect). However, it is hard to explain the observed concentration dependence purely in

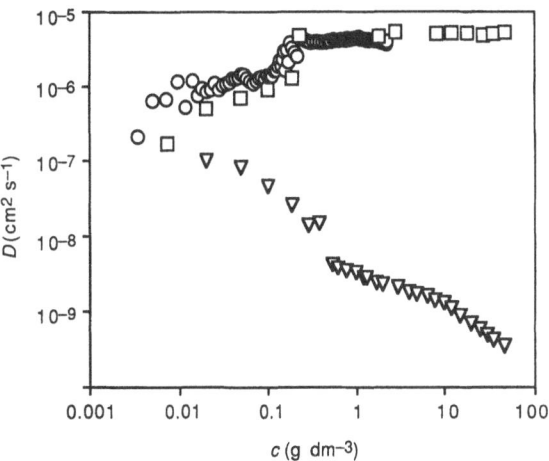

Fig. 5. Comparison on logarithmic axes of diffusion coefficient determined from synthetic boundary data for sodium poly(styrene sulphonate) ($M - 1.06 \times 10^6$) in salt-free solution ($\bigcirc$) with results from Ref. [34] for the fast ($\square$) and slow ($\triangledown$) diffusion coefficients determined by dynamic light scattering for a similar sample ($M = 1.2 \times 10^6$)

such terms. Genz et al. [35] suggest that both electrostatic repulsion and hydrodynamic interaction enhance the apparent diffusion coefficient. Another view [31] is that the fast diffusion coefficient represents the movement of polymer segments between junction points in a network structure (i.e. a gel mode of a transient network). For low molecular weight samples a constant fast diffusion coefficient seems to be reached at concentrations well below that at which semi-dilute solution behaviour would normally be expected [34]. However, long-range electrostatic interactions may give rise to an ordered lattice at very low concentrations. There is some evidence for the coexistence of ordered and disordered regions in salt-free polyelectrolyte solutions [8]. It certainly seems that interactions between polyions, as mediated by the counterions, and the consequent structure of polyelectrolyte solutions, cannot be ignored in the interpretation of the diffusion behaviour. It may be that it is best to view the fast diffusion process as being akin to the swelling of an electrostatically-stabilized pseudo-network.

The *slow* diffusion coefficient observed by dynamic light scattering also requires some explanation. It has been suggested [36] that it may be due to aggregates which are removable by filtration. Others [30–34] attribute it to the motion of electrostatically-stabilized multichain "domains".

More information is needed about the structure of salt-free polyelectrolyte solutions.

Viscosity

It is also of interest to compare the diffusion results with the viscosity behaviour of salt-free polyelectrolyte solutions. Figure 6 shows results obtained by Ise's group [5] for the reduced viscosity, at a shear rate of 100 s^{-1}, for sodium poly(styrene sulphonate) of similar molecular weight to that used in the diffusion study. On increasing the shear rate at a given concentration, the reduced viscosity decreases. Over the concentration range in which the diffusion coefficient is essentially constant, the reduced viscosity is seen to increase dramatically with dilution. At very low concentrations the reduced viscosity passes through a maximum (not shown in Fig. 6) [7]. The characteristic increase in reduced viscosity on dilution has in the past been attributed primarily to expansion of the polyion chain. However, qualitatively similar behaviour has been observed for charged latex particles, which cannot expand significantly [4]. It thus seems that factors other than conformational change need to be considered; intermolecular interactions must play a part.

Conclusions

The sedimentation and diffusion behaviour of polyelectrolytes in the presence of significant quantities of added salt is reasonably well understood. For the quantitative treatment of sedimentation data, one should take into account the influence of the Donnan effect on the apparent buoyancy of the macromolecules. This effect increases with increasing salt concentration and only disappears for the specific case of a salt with zero buoyancy factor.

The sedimentation and diffusion behaviour of polyelectrolytes in salt-free solution (or at very low salt concentration) raises many questions. In particular, theories of the "primary charge effect" or "coupled-ion diffusion" require critical reevaluation. A new approach to polyelectrolyte solutions is needed that takes into account the complex interactions of a polyion with its counterions and with other polyions.

Acknowledgement

The synthetic boundary diffusion results were obtained and patiently analysed by Miss Jane Wardle in the course of an undergraduate project.

References

1. Budd PM (1989) In: Allen G, Bevington JC, Booth C, Price C (eds) Comprehensive Polymer Science, Vol 1, Pergamon, Oxford, Chapter 10, p 199
2. Budd PM (1989) In: Allen G, Bevington JC, Booth C, Price C (eds) Comprehensive Polymer Science, Vol 1, Pergamon, Oxford, Chapter 11, p 215
3. Budd PM (1992) In: Harding SE, Rowe AJ, Horton JC (eds) Analytical Ultracentrifugation in Biochemistry and Polymer Science, Royal Society of Chemistry, Cambridge, Chapter 32
4. Yamanaka J, Matsuoka H, Kitano H, Ise N (1990) J Coll Interf Sci 134:92
5. Yamanaka J, Matsuoka H, Kitano H, Hasegawa M, Ise N (1990) J Am Chem Soc 112:587
6. Yamanaka J, Araie H, Matsuoka H, Kitano H, Ise N, Yamaguchi T, Saeki S, Tsubokawa M (1991) Macromolecules 24:3206
7. Yamanaka J, Araie H, Matsuoka H, Kitano H, Ise N, Yamaguchi T, Saeki S, Tsubokawa M (1991) Macromolecules 24:6156
8. Ise N (1985) Macromol Chem Suppl 12:215
9. Donnan FG (1911) Z Elektrochem 17:572
10. Casassa EF, Eisenberg H (1964) Adv Protein Chem 19:287
11. Eisenberg H (1976) Biophys Chem 5:243
12. Eisenberg H (1976) Biological Macromolecules and Polyelectrolytes in Solution, Oxford University Press, Oxford

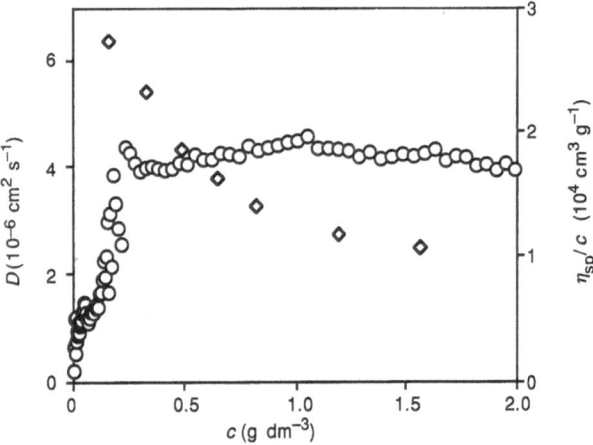

Fig. 6. Comparison of diffusion behaviour of sodium poly(styrene sulphonate) ($M = 1.06 \times 10^6$) in salt-free solution ($\bigcirc$) with results from Ref. [5] for the reduced viscosity of a similar sample ($M = 1.1 \times 10^6$) at a shear rate of 100 s^{-1} ($\diamond$)

13. Budd PM (1988) Brit Polym J 20:33
14. Budd PM (1985) Polymer 26:1519
15. Pederson KO (1958) J Phys Chem 62:1282
16. Mijnlieff PF (1963) In: Williams JW (ed), Ultracentrifugal Analysis, Academic, New York, p. 81
17. Varoqui R, Schmitt A (1972) Biopolymers 11:1119
18. Alexandrowicz Z, Daniel E (1963) Biopolymers 1:447
19. Wales M, Van Holde KE (1954) J Polym Sci 14:81
20. Berne BJ, Pecora R (1976) Dynamic Light Scattering, Wiley, New York, Chapter 9
21. Vink H (1990) J Coll Interf Sci 135:218
22. Säverborn S (1945) Dissertation, Uppsala
23. Nagasawa M, Fujita H (1964) J Am Chem Soc 86:3005
24. Crank J (1956) The Mathematics of Diffusion, Oxford University Press, Oxford
25. Boltzmann L (1894) Ann Physik, Leipzig 53:959
26. Matano C (1932–33) Jap J Phys. 8:109
27. Mandel M (1993) In: Brown W (ed), Dynamic Light Scattering: The Method and Some Applications, Oxford University Press, Oxford, Chapter 7
28. Lin SC, Lee WI, Schurr JM (1978) Biopolymers 17:1041
29. Schmitz KS, Lu M, Singh M, Ramsay DJ (1984) Biopolymers 23:1637
30. Sedlák M (1993) Macromolecules 26:1158
31. Förster S, Schmidt M, Antonietti M (1990) Polymer 31:781
32. Drifford M, Dalbiez JP (1985) Biopolymers 24:1501
33. Sedlák M, Amis EJ (1992) J Chem Phys 96:817
34. Sedlák M, Amis EJ (1992) J Chem Phys 96:826
35. Genz U, Benmouna M, Klein R (1991) Macromolecules 24:6413
36. Li X, Reed WF (1991) J Chem Phys 94:4568

Received June 11, 1993
accepted October 23, 1993

Author's address:

Dr Peter M. Budd,
Department of Chemistry,
University of Manchester,
Manchester M13 9PL, UK

Progress in Colloid & Polymer Science   Progr Colloid Polym Sci 94:116–119 (1994)

Computeranalysis of ultracentrifugation interference patterns

D. Steinmeier

University of Osnabrück, FRG

Abstract: A program for the computeranalysis of ultracentrifugation interference patterns is presented. The interference photographs are digitized using a HP Scanjet scanner or must be 1 bitmap .BMP files from other sources.

The program finds three consecutive fringes to calculate a "masterfringe" by averaging. Interruptions in the patterns are detected and substituted by self-finding unbroken fringes for averaging. The resolution in both directions depends on the resolution of the scan and is typically 10 μm (when scanning with 300 dpi). Output is a sequential file containing the calibrated x − y values of the masterfringe.

An application of the method to experimental data is shown.

Key words: Ultracentrifugation – interference – computer – scanner – pattern

Introduction

The purpose of the program is the evaluation of interference pattern obtained by ultracentrifugal measurements. Usually the interference patterns are digitized "by hand" using a digitizer: the coordinates of four reference points and the course of one fringe are transferred point by point; this procedure is time consuming and not very exact.

A program which could do the same job should be able to

1) load the complete interference pattern;
2) find the reference points by itself;
3) follow the exact course of a fringe.

Figure 1 shows a typical interference photograph of an ultracentrifugation experiment. For further calculations the exact coordinates of the four reference points (A-D) and the course of one fringe must be known. The four reference points are:

A) refb   reference (bottom of the cell);
B) refm   reference (meniscus of the cell);
C) rb   bottom of cell;
D) rm   meniscus of cell.

The other possibility to transfer an interference pattern to a computer is the scanning of the whole photograph. The main problem is now to find criteria which allow a safe recognition of the used reference points; furthermore, the exact course of a fringe must be followed without irritations, i.e., if a fringe has small interruptions.

The first step is the digitizing of the interference pattern photographs using a scanner (HP Scanjet II).

Scanning the photograph

Because the program can only handle 1-bitmap files, an important task is the suitable choice of the "cliplevel", that means the graylevel beyond which a point is interpreted as white and otherwise as black.

This can be done by varying the brightness and contrast adjustments of the scanner. The result should be a *b/w* - scan where fringes are clearly separated and only minor interruptions of fringes occur. The scans must be stored in Microsoft Windows .BMP format, which can be read directly by the program.

For good results a resolution of 200 dpi is sufficient and recommended; a higher resolution only takes a lot more storage capacity and

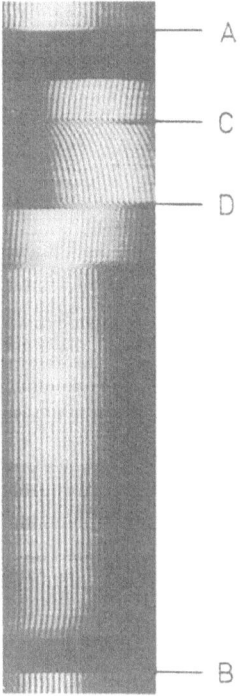

Fig. 1. Interference photograph of an ultracentrifugation experiment. The marked reference points are: A) reference (bottom of cell); B) reference (meniscus of cell); C) bottom of cell; D) meniscus of cell

calculation time but does not improve the results significantly.

After reading the .BMP file the program starts evaluation automatically.

Detection of reference points

As one can see in Fig. 1 the two reference points rm and rf are defined by the borderline between fringes and a completely dark region in the interference pattern. Therefore, the detection of these points is quite easy: the pattern is scanned horizontally line by line and the brightness is summarized for each line. When the dark region is entered or left, a sudden change of "linebrightness" should occur.

The fringes are parallel and nearly linear in regions, where no change in refractive index occurs. This is the case outside the solution. At the bottom of the cell and at the meniscus a strong change in refractive index causes a strong deflection of the fringes. The "linebrightness" seems to decrease in this region.

To test this, different interference pattern were linescanned.

Figure 2 shows as an example the "linebrightness" vs. scanline number. In the regions of the bottom of cell and meniscus, one can see an absolute minimum which corresponds to the real positions of the bottom of the cell and the meniscus.

Therefore, these two points should be found by linescanning and finding the absolute minimum of the linebrightness. Not all lines must be scanned for this purpose because the position of the bottom of the cell is nearly constant and the filling height of the cell normally does not vary so much. Therefore only certain plausible regions must be scanned.

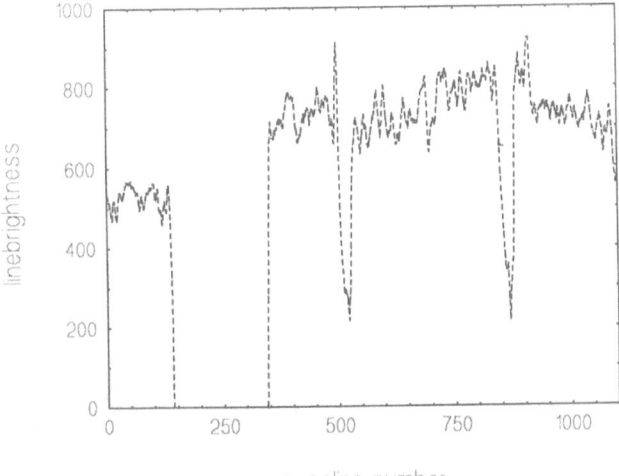

Fig. 2. Linebrightness (number of bright pixels of one scanline) vs. scanline number

Course of the fringes in the solution

To follow the course of the fringes inside the solution the program needs some interactive help from the user in this version of the program.

After finding all reference points the scanned region of the solution is automatically centered on the screen. The program can follow "dark" fringes. The user is requested to mark five points inside a "bright" fringe (using the mouse), which roughly follows the course of the fringe. These points are then connected linearly; the four resulting lines must be inside the region of the "bright" fringe.

If the program detects a collision of a line with a "dark" fringe, the input is rejected and the user is requested to repeat the procedure of marking points.

If the input is accepted, the four lines mark the starting points for the fringe detection. Starting from these points, the program tries to find three consecutive fringes, scanning the pattern horizontally in one direction. The two edges of every fringe are detected and the mean value of these coordinates is taken as the minimum of that fringe. The values for the three fringes are averaged to eliminate irregularities caused by the limited resolution of the scanner. The distance of the fringes is summarized and divided by the total number of fringe detections. In this way a very exact mean value for the distance of the fringes, which should be constant, is obtained.

If a fringe is interrupted, the program continues searching the next unbroken fringe; since the distance of the fringes is known, the obtained coordinates can be corrected to the "normal" course of the "original" fringe. If the end of the line is reached without finding three fringes, no values for this line are returned.

Because of the fringe deflection near the bottom of the cell and near the meniscus, these most important coordinates of the fringes cannot be obtained directly, but must be extrapolated. The quality of the extrapolation depends strongly on the number of known values; the program typically yields about 250–350 points per fringe, whereas a "human digitizer" can only pick up 20–40 points per fringe and must find the minimum of a fringe by visual estimation. Therefore, the automatic fringe detection should result in more precise extrapolated values and also in better results for the molecular weight distribution.

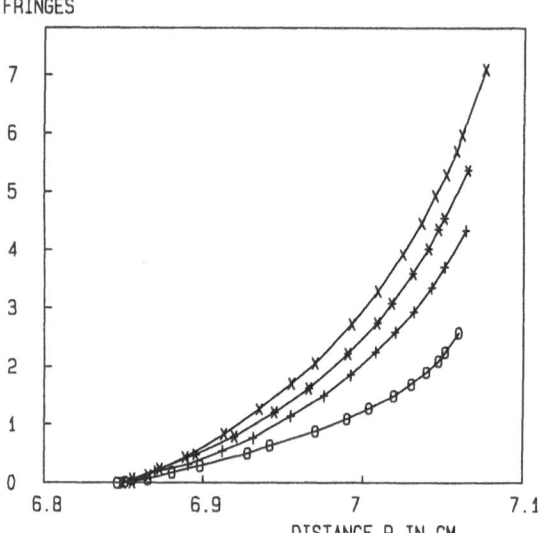

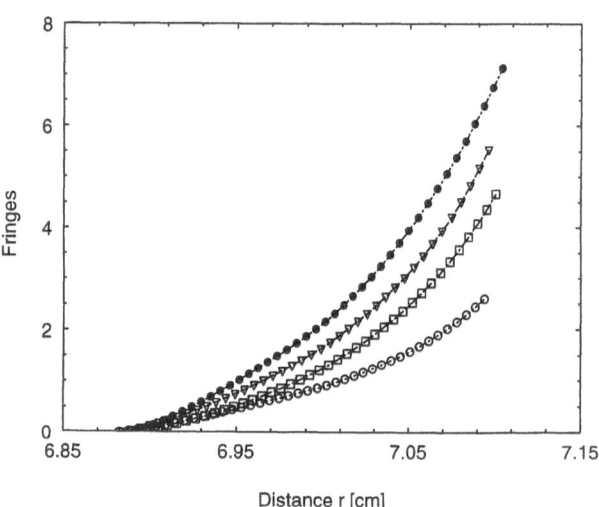

Fig. 3. Course of a fringe in the solution obtained by manual digitizing (a) and by the program (b) respectively

Results

The program was tested with a set of interference photographs which had been evaluated manually before.

The experiment was an equilibrium run of polystyrene in cyclohexane at four different concentrations between 0.5 and 2 g/l.

The calculations of M_{app}, $M_{w,app}$, and $M_{z,app}$ were performed using a method described by Lechner and Mächtle [1, 2].

Figure 3 shows the course of a fringe in the solution obtained by manual digitizing (a) and

Table 1. Comparison of the calculated molecular weights

	Comparison of the results					
	"by hand"			by computer		
$c/(g/l)$	M_{app}	$M_{w, app}$	$M_{z, app}$	M_{app}	$M_{w, app}$	$M_{z, app}$
0.5	260000	277200	380400	266800	281100	363600
1.02	216800	226700	314500	225900	245900	403900
1.51	180900	188700	307500	177600	197300	409700
1.99	177000	185100	307700	180200	192000	289200

by the program (b) respectively. Figure 3b only shows every 5th data point; for further calculations all data points are used.

The course of the fringe which was obtained by the program is much smoother than the manually digitized course. The program has objective criteria to find the maximum of a fringe: the detected edges of a fringe represent points of the same cliplevel, that means same optical density; since the optical density of a fringe is symmetrical with respect to the maximum, the mean value of two edges should result in a precise value for the maximum of the fringe.

The "human digitizer" has to estimate the position of the maximum for every point, which results in some jitter of the data points.

Figure 3 shows the course of the fringes including the extrapolated course at the bottom of the cell and at the meniscus for both methods. Because the program yields a lot of data points without any jitter, these extrapolations are much better than those obtained from manually digitized data points.

The extrapolated slopes at rm and rb are most important for the calculation of the different molecular weights. Table 1 shows the comparison of the calculated molecular weights. The deviations are quite small, but significant.

As explained above, we think that the results obtained by the program are more precise than those obtained by manual digitizing.

Using a special adapted hardware (i.e., CCD chip inside the ultracentrifuge) the data capture could be done directly without photographing the fringes. This should additionally improve the handling and the results.

References

1. Lechner MD, Mächtle W (1991) Makromol Chem **192**:1183–1192
2. Lechner MD, Mächtle W (1991) Progr Colloid Polym Sci **86**:62–69

Received June 23, 1993;
accepted September 23, 1993

Authors' address:

Dr. D. Steinmeier
Universität Osnabrück
Physikalische Chemie
Barbarastr.7
49069 Osnabrück

Author Index

Anderson AL   74

Behlke J   40
Borchard W   82, 90
Budd PM   107

Cölfen H   90

Davis SS   66
Demaine PD   74
Durchschlag H   20

Fiebrig I   66

Germeroth L   14

Harding SE   54, 66

Jungblut A   102

Kim SJ   46
Knespel A   40

Laue TM   74
Lewis MS   46

Michel H   14

Ortlepp B   102

Panke D   102

Ristau O   40

Schubert D   14
Schuck P   1, 14
Shrager R   46
Steinmeier D   116

Tziatzios C   14

van den Broek JA   14

Zipper P   20

Subject Index

ab initio calculation   20
analytical centrifugation   1
analytical centrifuge   102
analytical ultracentrifugation   40, 66,
  90
analytical ultracentrifuge   74, 107
aqueous solution   20
association constants   40

bioadhesion   66
biochemical model compounds   20

centrifugal field   82
chitosan   66
complex formation   40
computer   116
computerized Schlieren pattern
  evaluation   90
continuous equilibrium   82

diffusion   107
drug delivery   66
dynamic light scattering   107

equilibrium sedimentation   74
exponential fitting   1

gels   90

heterogeneous associations   1
hydrodynamics   74

instrumentation   74
interference   116
interferometry   74
ionic compounds   20

light scattering   102
light-harvesting complex
  B800/820   14

methods   74
molar mass   14
– averages   54
– distributions   54, 102
molecular interactions   46
mucins   54, 66

nonlinear least-squares
  techniques   40
nucleic acids   46

organic compounds   20

partial molar volume   20
– specific volume   14, 20
pattern   116

poly(methylmethacrylate)   102
polyelectrolyte   107
polymers   20
polysaccharides   54
protein assemblies   40
–/pigment/detergent micelles   14
proteins   46

Rayleigh optics   54
relation of other techniques   54

scanner   116
Schlieren optics   54
sedimentation analysis   102
– equilibrium   1, 14, 90
– velocity   107
size exclusion chromatography   102
sodium poly(styrene sulphonate)   107
swelling   82
– pressure   82, 90

ternary gel   82
thermodynamics   74
theta system   102

ultracentrifugal analysis   46
ultracentrifugation   116
UV scanner   102

velocity sedimentation   74